v a d e m e c u m

T0261908

Dermatopathology

Ramón L. Sánchez, M.D.
The University of Texas Medical Branch at Galveston
Galveston, Texas, U.S.A.

Sharon S. Raimer, M.D.
The University of Texas Medical Branch at Galveston
Galveston, Texas, U.S.A.

CRC Press
Taylor & Francis Group
Boca Raton London New York

CRC Press is an imprint of the
Taylor & Francis Group, an **informa** business

VADEMECUM
Dermatopathology

First published 2001 by Landes Bioscience

Published 2018 by CRC Press
Taylor & Francis Group
6000 Broken Sound Parkway NW, Suite 300
Boca Raton, FL 33487-2742

© 2001 by Taylor & Francis Group, LLC
CRC Press is an imprint of Taylor & Francis Group, an Informa business

No claim to original U.S. Government works

ISBN 13: 978-1-57059-496-0 (pbk)

Visit the Taylor & Francis Web site at
http://www.taylorandfrancis.com

and the CRC Press Web site at
http://www.crcpress.com

Library of Congress Cataloging-in-Publication Data
Dermatopathology / [edited by] Ramón Sánchez, Sharon Raimer.
    p.;cm.--(Vademecum)
    Includes index.
ISBN 1-57059-496-1
    1. Skin--Diseases-Handbooks, manuals, etc. I. Sanchez, Ramon L. II. Raimer, Sharon. III. Series.
    [DNLM: 1. Skin Diseases--pathology. WR 140 D435172 2001]
RL96.D476 2001
616.5--dc21   01-042377

Dedication

To the students, residents and fellows through whom we learn so much in the process of teaching.

To our families, always supportive of our endeavors.

Contents

Editors

Ramón L. Sánchez, M.D.
Clinical Professor
Departments of Dermatology and Pathology
The University of Texas Medical Branch at Galveston
Galveston, Texas, U.S.A.
Chapters 4-9, 12-17, 19-22

Sharon Raimer, M.D.
Professor and Chairman
Department of Dermatology
The University of Texas Medical Branch at Galveston
Galveston, Texas, U.S.A.
Chapters 4-9, 12-17, 19-22

Contributors

San-Hwan Chen
Department of Dermatology
The Everett Clinic
Everett, Washington, U.S.A.
Chapters 11 and 23

Marcia Driscoll
Associates in Dermatology
Orlando, Florida, U.S.A.
Chapter 2

Daniel B. Crump
Keesler Medical Center
Keesler AFB, Mississippi, U.S.A.
Appendix

Gina Harney
Texoma Dermatology Clinic, P.A.
Sherman, Texas, U.S.A.
Chapter 1

John R. Simmonds
Pathology Associates, Inc., P.S.
Spokane, Washington, U.S.A.
Chapter 10

Gayle S. Westhoven
Associated Dermatologists, P.A.
Birmingham, Alabama, U.S.A.
Chapter 3

Angela Yen Moore
Clinical Assistant Professor
The University of Texas Medical Branch
Galveston, Texas, U.S.A.
Chapter 18

Preface

This book is a primer in dermatopathology, directed to residents of pathology and dermatology, and to medical students with an interest in dermatology. Since the contents are quite inclusive, with some 500 entities described or mentioned, it can be of help also to practicing dermatologists, general pathologists, and primary care physicians who come across skin lesions and/or skin biopsies in their practices. This book does not intend to be a reference text of dermatopathology, hence the lack of references at the end of the chapters. The interested reader should consult other textbooks of dermatopathology or the current medical literature for further information on the subjects.

Our intention has been to write a very practical manual, and to organize and present its contents with the beginner student of dermatopathology in mind. Each section starts with the definition of a basic morphologic change under which several dermatopathology conditions are grouped. In order to avoid duplication, each entity is described only once in the section most characteristic of its histologic findings. However, many conditions are cited also in other sections describing changes that pertain also to those entities. For instance, psoriasis is described under parakeratosis, but is also cited under subcorneal microabscesses and under acanthosis. This form of presenting information results indeed in the formation of a differential diagnosis for each histologic finding. Furthermore, we tried to construct lists of meaningful differential diagnoses, including only entities that characteristically relate to the particular histologic change, rather than construct long lists that lack specificity.

We hope the reader will find this manual as practical and useful as we intended it to be. If we succeeded even in part to achieve those goals, our efforts will have been fully compensated.

The Editors

Acknowledgments

Susan Keefer transcribed the original dictations and played an important role in having this work initiated. Pat Perez and Marian Kimbrough typed and proofread the manuscript.

CHAPTER 1

Hyperkeratosis and Parakeratosis

Gina Harney

Section 1. Hyperkeratosis

Hyperkeratosis implies increased thickening of the stratum corneum. The normal appearance of this layer is the "basket weave" appearance, or layered keratin that is loosely attached. In hyperkeratosis the stratum corneum usually appears solid, in addition to thick. Synonym: Orthokeratosis.

Clinically the hyperkeratotic processes are manifested by scales that peel-off with relative ease. In the case of keratodermas, the thickening of the stratum corneum may be extraordinary.

Ichthyosis Vulgaris

Clinical

An autosomal dominant condition with dry skin to fish- like scales, most prominent on extensor extremities. Presents before age 5. Associated with keratosis pilaris and atopy. Represents a retention hyperkeratosis associated with a defect in filaggrin synthesis.

Histology

Histology shows hyperkeratosis with a thin or absent granular layer. This histologic finding is almost unique, since a granular layer is almost always present under a hyperkeratotic stratum corneum (Fig. 1.1).

X-Linked Ichthyosis

Clinical

Since the disease is X-linked recessive, males have the more severe form. Large scales thicken with age, giving "dirty" appearance. Corneal opacities and cryptorchidism are associated. Deficiency of arylsulfatase C or steroid sulfatase underlies this retention hyperkeratosis.

Histology

Histology reveals hyperkeratosis with a normal or slightly thickened granular layer.

Table 1.1. Hyperkeratotic processes

Ichthyosis vulgaris
X-linked ichthyosis
Lamellar ichthyosis
Poikiloderma
Keratoderma
Pachyonichia congenita
Flegel's disease
Inflammatory linear epidermal nevus
Vitamin A deficiency
Discoid lupus erythematosus
Lichen sclerosus et atrophicus
Lichen planus

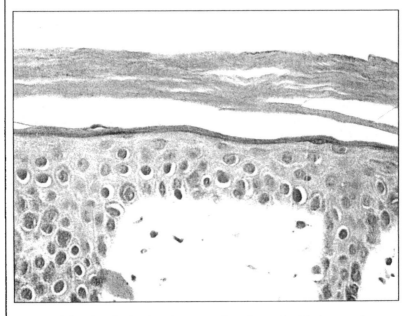

Fig.1.1. **Ichthyosis vulgaris**: There is compact hyperkeratosis with absence of granular layer.

Lamellar Ichthyosis

Clinical

An autosomal recessive condition which may present as a collodion baby. Large, extensive, plate-like scales are present frequently with ectropion, possibly erythroderma in this retention hyperkeratosis.

Histology

Histology shows hyperkeratosis, granular layer present, often with psoriasiform hyperplasia.

Palmoplantar Keratoderma

Clinical

Several types have been described. The **Autosomal Dominant Unna-Thost** form (orthohyperplasia) and the **Vorner** form (epidermolytic hyperkeratosis) have an onset in infancy of sharply marginated, diffuse palmoplantar hyperkeratosis. Keratin 9 mutations have been described. In the **punctate** form numerous palmoplantar keratotic plugs are present.

Autosomal recessive Mal de Meleda has an onset in early infancy. Hyperkeratosis spreads to dorsal surfaces (transgrediens) and nail dystrophy and pseudoainhum are present. **Papillon-Lefevre** syndrome also shows transgediens, with periodontopathy.

Palmoplantar keratoderma is also seen in pachonychia congenita, hidrotic ectodermal dysplasia, tyrosinemia type II, Vohwinkel syndrome, Olmsted's syndrome, dyskeratosis congentia, epidermolysis bullosa simplex, Howell-Evans syndrome, eccrine syringofibroadenomatosis, and as a presentation of cutaneous T-cell lymphoma.

Histology

Histologically, all forms of keratoderma reveal ortho-hyperkeratosis, hypergranulosis, and acanthosis except the Vorner form in which epidermolytic hyperkeratosis, characterized by vacuolation of stratum malpighii cells forming cavities due to cell wall rupture, is seen (Fig.1.2). The punctate lesions are sharply delimited.

Pachyonychia Congenita

This condition is autosomal dominant with an onset in childhood of palmoplantar keratoderma with painful callosities over pressure points, subungueal hyperkeratosis, oral mucosal leukokeratosis, and follicular hyperkeratosis. Keratin 16 and 17 defects have been described. Histologically there is orthohyperkeratosis. Additionally, blisters may arise in upper stratum malpighii.

Flegel's Disease (Hyperkeratosis Lenticularis Perstans)

Clinical

Adult onset of numerous persistent flat, scaly papules up to 5mm in diameter. Most prominent on dorsa of feet and lower legs.

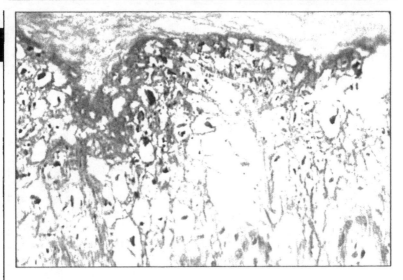

Fig.1.2. **Palmoplantar keratoderma/epidermolytic hyperkeratosis**: There is hyperkeratosis and acanthosis. The keratinocytes appear vacuolated, empty, and there is an increased granular layer.

Histology

Histology shows hyperkeratosis, occasional foci of parakeratosis, epidermal flattening with a decreased granular layer and a lichenoid or perivascular chronic papillary dermal infiltrate.

Vitamin A Deficiency (Phrynoderma)

Clinical

Occurs mainly in Asia and Africa, also associated with malabsorption. Leads to dry, rough skin, follicular hyperkeratosis and slow wound healing. Also causes night blindness, xerophthalmia and keratomalacia.

Histology

Histologically, hyperkeratosis with horny plugs in hair follicles as well as sebaceous and sweat gland atrophy is seen.

Section 2. Parakeratosis

Parakeratosis implies preservation of picnotic, flat nuclei in the stratum corneum (which is usually anucleated). Parakeratosis can be classified as confluent, when there is extensive, continous parakeratosis, or columnar when only focal, vertical columns of parakeratosis are present. The stratum of Malpighii under an area of parakeratosis usually lacks a granular layer.

Table 1.2. Confluent parakeratosis

Psoriasis
Reiter's disease
Chronic dermatitis
Seborrheic dermatitis
Pityriasis lichenoides

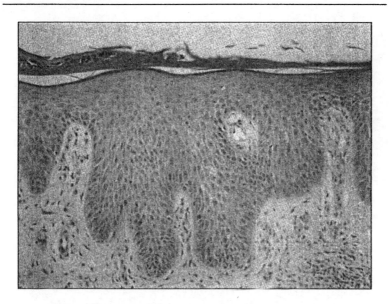

Fig.1.3. **Psoriasis**: Confluent parakeratosis and regular acanthosis.

Confluent Parakeratosis

Psoriasis

Clinical

Psoriasis is a chronic, fluctuating inflammatory skin disease. The details of inheritance and the role of environmental factors remain to be elucidated. Many HLA antigens have been associated with psoriasis. Of these, HLA Cw6 is felt to have the strongest association. A spectrum of clinical presentations exist: plaque type (most common), guttate, inverse, pustular (generalized or localized), and erythroderma. The characteristic plaque of psoriasis is well-circumscribed with silvery scale. Several variants of pustular psoriasis have been described, including **Von Zumbusch, impetigo herpetiformis** and **acrokeratosis continua of Hallopeau.**

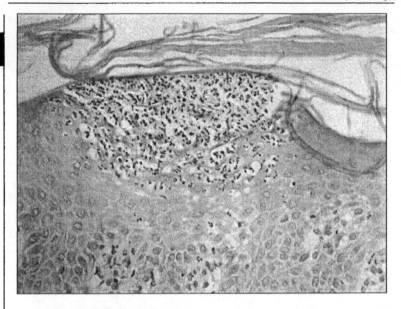

Fig.1.4. **Spongiform pustules/Munro microabscesses**: A collection of neutrophils is seen in the upper part of the stratum of Malpighii.

Fig.1.5. **Pustular psoriasis**: A large subcorneal pustule which involves also the external half of the stratum of Malpighii.

1

Histology

The epidermis is acanthotic (regular acanthosis) with elongated, "club shaped" rete ridges. The granular layer is absent and extensive, confluent parakeratosis and thinning of the suprapapillary plate are seen (Fig. 1.3). Spongiform pustules of Kogoj represent collections of neutrophils which have migrated into the upper part of the stratum of Malpighii (Fig. 1.4). These become Munro microabscesses as these collections reach the corneal layer. In the papillary dermis, increased number of dilated capillaries and edema are present. In pustular psoriasis, the pustules of Kogoj occur as macropustules which may appear intraepidermal (Fig. 1.5). Biopsies of psoriatic plaques cannot be differentiated from subacute or chronic dermatitis if Munro microabscesses or spongiform pustules are not present. In that situation they are often diagnosed as "psoriasiform dermatitis".

Reiter's Disease

Clinical

Most commonly affects young men, most of whom express HLA-B27 antigen. The syndrome is a triad of urethritis, arthritis and conjunctivitis following chlamydia or other associated bacterial infections. Cutaneous findings include balanitis circinata (a superficial ulceration of the glans penis) and keratoderma blennorrhagicum (erythematous hyperkeratotic plaques, most frequently palmoplantar).

Histology

Histologically, difficult to distinguish from psoriasis with psoriasiform epidermal hyperplasia, parakeratosis, spongiform pustules in the epidermis and papillary dermal edema and dilated capillaries.

Chronic Dermatitis

Clinical

Chronic dermatitis is a chronic inflammatory process in the skin which may be clinically diagnosed as contact dermatitis, atopic dermatits, lichen simplex chronicus, seborrheic dermatitis or stasis dermatitis.

Histology

Hyperkeratosis with areas of parakeratosis, often confluent, overlying an acanthotic epidermis with elongation of the rete ridges. Slight spongiosis may be present, although much less than in acute dermatitis (Fig. 1.6). The upper dermis shows a lymphocytic perivascular inflammatory infiltrate with capillary proliferation and occasional mild edema. Dermal hemosiderin is an additional finding in stasis dermatitis.

Seborrheic Dermatitis

Clinical

A common chronic condition which in adults presents as erythema, pruritus and scaling, most prominently involving scalp, ears, browline and nasolabial folds.

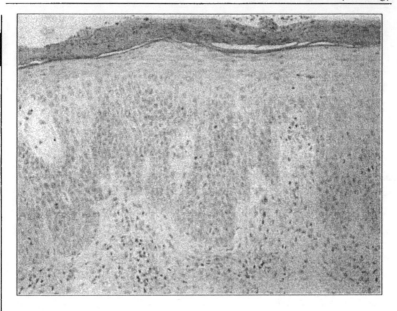

Fig.1.6. **Chronic dermatitis/lichen simplex chronicus**: Parakeratosis and irregular acanthosis. Occasional spongiosis.

Table 1.3. Parakeratosis: columnar or focal

Pityriasis rosea
Pityriasis rubra pilaris
Porokeratosis

Infantile presentation often also involves the diaper area. Extensive involvement is common in AIDS patients.

Histology

Histology shows hyperkeratosis with foci of parakeratosis and occasional pyknotic neutrophils in the stratum corneum, similar to Munro microabscesses seen in psoriasis. The epidermis is acanthotic with elongated rete ridges. Mild spongiosis is typical. The upper dermis reveals a mild chronic inflammatory infiltrate.

Parakeratosis: Columnar or Focal

Pityriasis Rosea

Clinical

Pityriasis rosea is a self-limited papulosquamous eruption of unknown etiology. The majority of the patients are between 10 and 35 years of age. A primary lesion, the "herald patch", proceeds a generalized, symmetric eruption of oval, erythema-

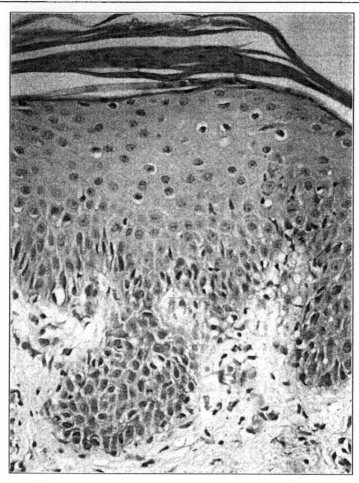

Fig.1.7. **Pityriasis rosea**: Columnar parakeratosis, focal spongiosis, and extravasation of red blood cells.

tous plaques with a fine collarette of scale on the trunk and proximal extremities. This eruption spreads, then fades over approximately six weeks.

Histology

Hyperkeratosis with focal, columnar parakeratosis overlying an acanthotic and spongiotic epidermis is seen. The presence of dyskeratotic, eosinophilic keratinocytes in the epidermis and of extravasated erythrocytes in the dermal papillae distinguishes pityriasis rosea from dermatitis. A lymphocytic perivascular infiltrate is seen in the upper dermis, which may migrate into the epidermis (Fig. 1.7).

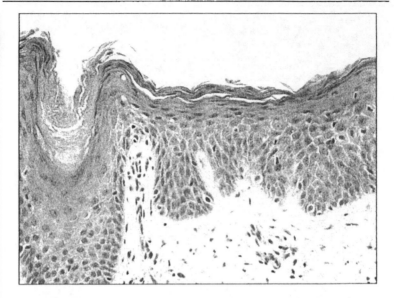

Fig.1.8. **Pityriasis rubra pilaris**: Alternating parakeratosis and orthokeratosis. Hair follicles show shoulder parakeratosis.

Pityriasis Rubra Pilaris

Clinical

Pityriasis rubra pilaris is an erythematous papulosquamous disorder of unknown etiology. Age of onset peaks bimodally, in both the first and fifth decade. The classic adult presentation accounts for over 50% of patients. Follicular papules with perifollicular erythema, that progress to become confluent and sometimes erythrodermic, leaving well-demarcated patches of normal skin (islands of sparing). Additionally, follicular plugging, palmoplantar keratosis with an orange hue, nail plate thickening and ectropion can occur.

Histology

Keratotic follicular plugs, hyperkeratosis alternating with foci of parakeratosis, parakeratosis around the follicular ostia, uniform epidermal acanthosis with broad epidermal ridges, and hypergranulosis distinguishes pityriasis rubra pilaris. Additionally, focal acantholysis with/without dyskeratosis has also been described. The dermis shows a chronic perivascular infiltrate (Fig. 1.8).

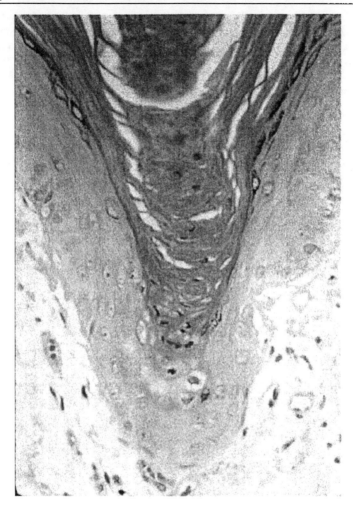

Fig.1.9. **Porokeratosis**: Columnar lamella is characteristic. Underneath part of the stratum of Malpighii appears vacuolated and necrotic.

Porokeratosis

Clinical

Primarily an autosomal dominant inheritance. At least four clinical forms have been described. **Porokeratosis of Mibelli** includes one or more plaque-type hyperkeratotic plaques which usually present in the first decade, and can be linear in distribution. **Porokeratosis palmaris, plantaris, et disseminata** often begins in young adults on palms and soles, spreading to form uncountable small plaques over the neck, trunk and extremities. **Disseminated superficial actinic porokeratosis** present most commonly on extensor extremities in the third to fourth decade. **Punctate porokeratosis** are 1-2 mm keratotic plugs limited to palmar and plantar surfaces. All lesions of porokeratosis except the punctate form show a peripheral "thread-like" keratotic border. Malignant degeneration to squamous cell carcinoma has been observed in porokeratosis.

Histology

The **cornoid lamella** is the hallmark of porokeratosis. Most prominent in the Mibelli type, the cornoid lamella is found along the outer border of the lesions. A keratin-filled epidermal invagination reveals a vertical tier of parakeratosis. The epidermis at the base of the parakeratotic column lacks a granular layer and contains vacuolated or pyknotic epidermal cells. The epidermis of the remainder of the lesion has a normal granular layer and occasional dyskeratotic or dysplastic cells are found in the epidermis. A vacuolar interface dermatitis is sometimes seen in the epidermis of the center of the lesion (Fig. 1.9).

CHAPTER 2

Acanthosis

Marcia Driscoll

Section 3. Acanthosis

Acanthosis implies hyperplasia or increased thickening of the stratum Malpighii. Usually the rete ridges are preserved, elongated and/or thickened. The suprapapillary portions of the epidermis may appear thinned (psoriasis), normal or thickened. Acanthosis is subdivided into regular, when the rete ridges end at about the same level, and irregular. We used the term lacelike acanthosis when there is elongation of rete ridges without thickening of epidermis.

Regular Acanthosis

Psoriasiform Dermatitis

There are multiple disorders which histologically may demonstrate regular acanthosis similar to that seen in psoriasis, but lack all of the classic features of psoriasis. The following is a list of such disorders, with features that differ from psoriasis:

1. Chronic eczematous dermatitis—often acanthosis is more irregular than that of psoriasis
2. Mycosis fungoides—epidermotropism of atypical lymphocytes, which may form Pautrier microabscesses
3. Parapsoriasis—similar to mycosis fungoides except cytologic atypia not definitively identified
4. Pityriasis rubra pilaris—classically has perifollicular parakeratosis
5. Reiter's disease—early lesions similar to psoriasis. Older lesions may be distinguished by massive hyperkeratosis
6. Bowen's disease—"windblown" appearance of atypical keratinocytes
7. Secondary syphilis—perivascular infiltrate may include plasma cells, dermal blood vessels may be dilated with large endothelial cells
8. Pityriasis rosea—dyskeratotic keratinocytes, extravasated red blood cells in dermal papillae, columnar parakeratosis.
9. Incontinentia pigmenti—eosinophils within epidermis

Inflammatory Linear Verrucous Epidermal Nevus

Clinical

Persistent linear array of erythematous, verrucous papules that most often arises in early childhood on the lower leg and/or adjacent hip or buttock.

Dermatopathology, edited by Ramón L. Sánchez and Sharon S. Raimer. ©2001 Landes Bioscience.

2

Table 2.1. Differential Diagnosis

Psoriasis
Psoriasiform dermatitis
Reiter's disease
Inflammatory linear epidermal nevus (ILVEN)

Histology
Regular acanthosis, slight spongiosis, alternation of parakeratotic areas with hypogranulosis and hyperkeratotic areas with presence of granular layer. Chronic inflammatory infiltrate in dermis (Fig. 2.1).

Irregular Acanthosis

Prurigo Simplex/Prurigo Nodularis

Clinical
Prurigo simplex consists of intensely pruritic erythematous papules, often on the extensor aspects of the arms. The lesions of prurigo nodularis consist of dome-shaped nodules that may be eroded or crusted, and are due to persistent rubbing or scratch

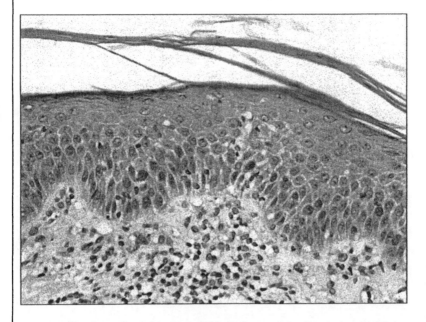

Fig.2.1. **Inflammatory linear verrucous epidermal nevus**: Hyperkeratosis, acanthosis, spongiosis, and perivascular inflammation.

Table 2.2. Irregular acanthosis

Lichen planus
Prurigo
Lichen simplex chronicus (chronic dermatitis)
Pemphigus vegetans
Blastomycosis, chromomycosis
Blastomycosis-like pyoderma
Bromoderma/iododerma
Pseudo epitheliomatous hyperplasia
Acanthoma fissuratum

---

ing. These lesions may be idiopathic or associated with a cutaneous or systemic disorder that causes pruritus.

Histology

Prurigo simplex has the appearance of a subacute dermatitis, with mild irregular acanthosis, spongiosis, spongiotic vesicles, parakeratosis, and a superficial perivascular chronic inflammatory infiltrate. Prurigo nodularis has more marked irregular acanthosis, often approaching pseudocarcinomatous hyperplasia. The architecture is that of a papule, and there may be progressive elongation of the rete ridges from the edge to the center of the lesion. In addition, there is hyperkeratosis with focal parakeratosis, hypergranulosis, a perivascular chronic inflammatory infiltrate, and papillary dermal fibrosis. Hypertrophy of nerve bundles may be noted (Fig. 2.2).

Differential Diagnosis

The clinical differential diagnosis of prurigo simplex includes arthropod bites, scabies, folliculitis, papular urticaria, papular eczema, miliaria rubra, dermatitis herpetiformis, and Grover's disease. The clinical differential diagnosis of prurigo nodularis includes nodular scabies, hypertrophic lichen planus, and persistent arthropod bite reactions. Histologically prurigo simplex may be difficult to distinguish from papular urticaria; prurigo nodularis is similar to other disorders that show the histologic pattern of chronic eczematous dermatitis, such as atopic dermatitis, nummular eczema, chronic contact dermatitis, and pityriasis rubra pilaris.

Lichen Simplex Chronicus (Chronic Dermatitis)

Clinical

Pruritic, well-circumscribed, lichenified plaque, often located on posterior neck or ankles, produced by chronic rubbing. It may be idiopathic or associated with a cutaneous or systemic disorder causing pruritis.

Histology

Irregular acanthosis, hyperkeratosis, hypergranulosis, chronic perivascular inflammatory infiltrate, and fibrosis of papillary dermis are present (see Fig. 1.6).

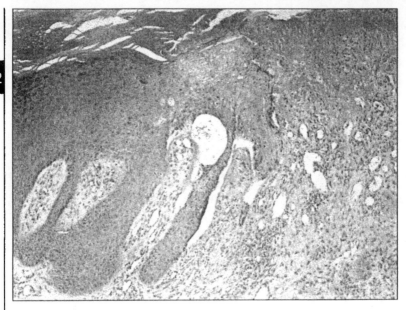

Fig.2.2. **Prurigo**: Irregular acanthosis and hyperkeratosis. An excoriation is present on the right side of the photograph.

Pseudoepitheliomatous (Pseudocarcinomatous) Hyperplasia

Clinical

This term is used to describe marked acanthosis that may be similar to well-differentiated squamous cell carcinoma. The following list includes cutaneous lesions which may display this characteristic:

1. Bromoderma, iododerma, associated with ingestion of bromides or iodides, can cause acneiform eruptions and intraepidermal and dermal abscesses.
2. Deep fungal infections—coccidiomycosis, blastomycosis, chromomycosis, paracoccidiomycosis (Chapter 14)
3. Pemphigus vegetans
4. Prurigo nodularis
5. Orf and milker's nodule
6. Keratoacanthoma
7. Hypertrophic lichen planus
8. Condyloma acuminata
9. Granular cell tumor
10. Tuberculosis verrucosa cutis
11. Acanthoma fissuratum

2

Acanthoma Fissuratum

Clinical
Nodule or plaque with a central fissure along postauricular creases or at the nasal bridge due to pressure of eyeglass frames or earpieces.

Histology
Marked acanthosis which may approach pseudocarcinomatous hyperplasia, depressed area of atrophic epidermis or epidermal separation filled with degenerated collagen (corresponding to location of fissure), and hyalinized collagen below the fissure.

Lacelike Acanthosis

Acanthosis Nigricans

Clinical
Hyperpigmented velvety thickening of skin that may occur on the posterior neck, axillae, inter- and inframammary area, groin, and antecubital or popliteal fossae. It is commonly associated with insulin resistant diabetes and hyperinsulinemia. It is most often associated with obesity and can arise at any age. It may also occur as part of a syndrome that includes hyperandrogenemia and hyperinsulinemia ("HAIR-AN syndrome"), as well as part of a variety of other insulin-resistant states. If the onset of acanthosis nigricans is sudden with rapid spread of skin lesions, it may be a marker for an underlying gastric adenocarcinoma.

Histology
Mild lacy acanthosis, hyperkeratosis, papillomatosis are present. There may be hyperpigmentation of the basal layer, but more often the cutaneous hyperpigmentation is secondary to hyperkeratosis (Fig. 2.3).

Differential Diagnosis
The clinical differential diagnosis includes confluent and reticulated papillomatosis (Gougerot-Carteaud), Dowling-Degos disease, and ichthyosis hystrix. The histologic differential diagnosis includes other disorders with papillomatosis. Confluent and reticulated papillomatosis shows milder hyperkeratosis and papillomatosis than acanthosis nigricans. Epidermal nevi may show greater acanthosis. Seborrheic keratoses of the hyperkeratotic type may show similar histology.

Confluent and Reticulated Papillomatosis (Gougerot-Carteaud)

Clinical
Slightly verrucous papules, occurring most often on the sternal region and in the midline of the back, that become confluent and form a reticulated pattern peripherally. It occurs more often in females, usually at or around the time of puberty. *Malassezia furfur* has been isolated from some lesions, but its role in this disorder is unclear.

Table 2.3. Differential diagnosis of lacelike acanthosis

Acanthosis nigricans
Confluent and reticulated papillomatosis of Gougerot and Carteaud
Seborrheic keratosis
Dowling-Degos (Reticulated pigmented dermatosis of the flexures)

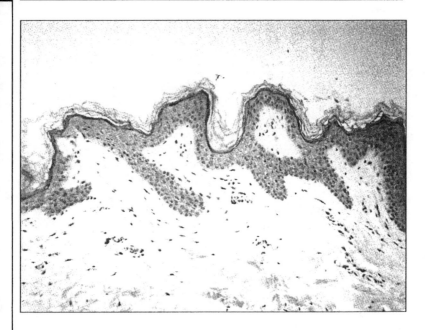

Fig.2.3. **Acanthosis nigricans**: Irregular acanthosis and papulomatosis.

Histology
Mild hyperkeratosis and papillomatosis with acanthosis limited to areas between dermal papillae (Fig. 2.4).

Dowling-Degos Disease (Reticulate Pigmented Dermatosis of the Flexures)

Clinical
Smooth hyperpigmented reticulated macules, confluent centrally and discrete peripherally, occurring on the axillae, inframammary area, and groin. Onset is in adulthood and the condition progresses slowly.

Histology
Irregular acanthosis with thin, elongated, and markedly hyperpigmented rete ridges and downward proliferation from the infundibula of hair follicles (Fig. 2.5).

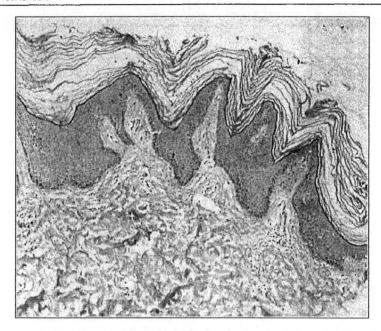

Fig.2.4. **Confluent and reticulated papillomatosis (Gougerot-Carteaud):** Hyperkeratosis, papillomatosis, and acanthosis. Pityrosporum organisms often present.

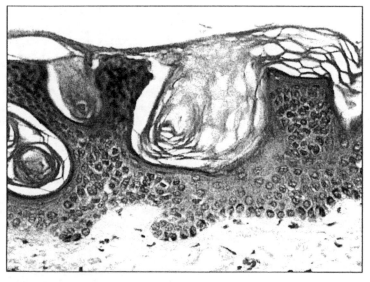

Fig.2.5. **Dowling-Degos disease:** Hyperkeratosis and acanthosis with basal melanosis.

CHAPTER 3

Papillomatosis, Atrophy and Alterations of the Granular Layer

Gayle S. Westhoven

Section 4. Papillomatosis

Papillomatosis results from the upward proliferation of dermal papillae causing the epidermis to be undulated. It results in a "warty" clinical appearance. It is often accompanied by hyperkeratosis, parakeratosis and acanthosis, but the hallmark of papillomatosis is the upward projections of the epidermis.

Verruca/Epidermodysplasia Verruciformis

Clinical

Verrucae present in a variety of clinical patterns, all caused by subtypes of the human papillomavirus (HPV), including verruca vulgaris (HPV-4, 7), palmoplantar wart (HPV-1 and -2), verruca plana (HPV-3), and condyloma acuminatum (HPV-6, 11, 16, and 18). Epidermodysplasia verruciformis (HPV-5, 8, 12, 14, 15 and 17) presents as extensive erythematous scaly macules or papules and seems to be associated with an autosomal recessive defect in cell mediated immunity. There is an increased incidence of skin cancer in long term lesions of epidermodysplasia verruciformis, especially in sun-exposed areas.

Histology

Verrucae vulgaris are characterized by compact type hyperkeratosis, acanthosis and papillomatosis. There are tiers of parakeratosis over the tips of the dermal papillae, as well as hypergranulosis and foci of koilocytotic cells (Fig. 3.1). Increased, eosinophilic keratohyalin may be seen. Findings vary depending on the clinical appearance and the age of the wart. Verrucae plana show hyperkeratosis and acanthosis but no papillomatosis. The stratum corneum has a basket weave appearance. In the upper stratum of malpighii there is a diffuse vacuolization of cells, some being enlarged to about twice their normal size. Deep palmoplantar (myrmecia) warts exhibit an exuberant amount of keratohyalin, which is eosinophilic rather than basophilic, resembling inclusion bodies (Fig. 3.2). In condyloma acuminatum the stratum corneum is only slightly thickened. Mucosal lesions show parakeratosis. The stratum malpighii shows papillomatosis and acanthosis with the rete ridges branching to such an extent that the lesions may resemble pseudocarcinomatous hyperplasia, resulting in cauliflower-like architecture (Fig. 3.3). Mitoses may be present. The most important feature for diagnosis is the presence of koilocytosis, in which epithelial cells in the

Table 3.1. Differential diagnosis of squamous papillomas

Verruca/epidermodysplasia verruciforme
Seborrheic keratosis
Hypertrophic actinic keratosis
Nevus sebaceous
Linear epidermal nevus
Acrokeratosis verruciformis of Hopf

3

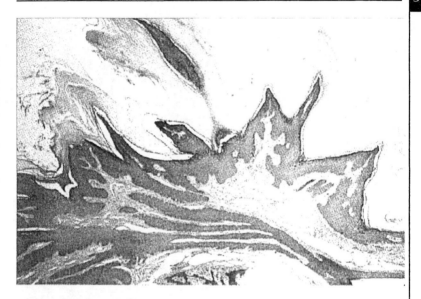

Fig. 3.1. **Verruca vulgaris**: Suprapapillary parakeratosis, marked papillomatosis, and acanthosis.

upper part of the stratum Malpighii show distinct perinuclear vacuolization and irregular, nonpicnotic nuclei. The dermis may show edema, dilated capillaries and a chronic inflammatory infiltrate. In epidermodysplasia verruciformis, the lesions typically display hyperkeratosis, acanthosis, and hypergranulosis. Keratinocytes may be swollen and pale with perinuclear vacuolization. With atypical lesions, these cells are larger and more hyperchromatic.

Nevus Sebaceous (Organoid Hamartoma)

Clinical

Nevus sebaceous is characteristically located on the scalp, but may be found anywhere on the head, neck or upper trunk. The lesion presents at birth as a well-circumscribed linear or oval slightly raised plaque. During late childhood and adolescence lesions enlarge and develop a verrucous appearance. Basal cell carcinoma and syringocystadenoma papilliferum as well as many other neoplasms may develop

3

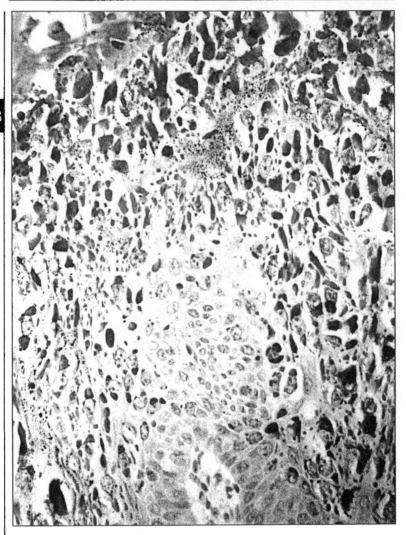

Fig.3.2. **Palmoplantar wart (myrmecia):** There is marked hyperplasia of keratohyalin granules in the stratum granulosum.

within a nevus sebaceous. Rarely, central nervous system symptoms such as seizures may be associated with these lesions.

Histology
The histology varies based on the developmental stage of the nevus sebaceous, with variable abnormalities of epidermis, hair follicles, sebaceous glands and apo-

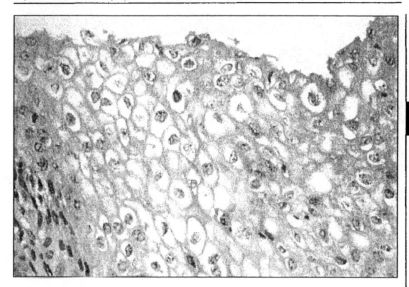

Fig.3.3. **Condyloma acuminatum**: There is vacuolization of keratinocytes with irregular nuclei, which are characteristic findings in koilocytosis.

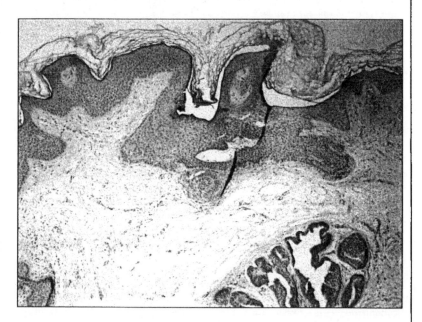

Fig.3.4. **Nevus sebaceous**: The epidermis shows papillomatosis and irregular acanthosis. Two follicular buds are connected to the epidermis.

crine glands. Early plaques are characterized by a decreased number of follicles, many of which are malformed, and underdeveloped sebaceous glands. In adolescence, the epidermis becomes papillomatous and irregular. Basaloid follicular germ nests are seen connected to the epidermis, often resembling superficial basal cell carcinoma. (Similar basaloid hyperplasia is also seen in dermatofibromas). Sebaceous glands are increased in number and/or in size, and are located high in the dermis. It is common for sebaceous glands in a nevus sebaceous to open directly in the epidermis, rather than in the hair follicle (Fig. 3.4). Apocrine glands showing characteristic decapitation secretion are present in the lower dermis up to two-thirds of the time. In adulthood, tumors may develop within the lesions, the two more common being syringocystadenoma papilliferum and basal cell carcinoma.

Differential Diagnosis

Differential diagnosis includes epidermal nevus, sebaceous hyperplasia and sebaceous adenoma.

Linear (Verrucous) Epidermal Nevus

Clinical

Linear epidermal nevi are usually present at birth but may develop in childhood. They present as verrucous yellow-brown papules in a linear arrangement and can be located anywhere on the body. The localized form consists of only a single linear lesion. The systemic form, however, consists of multiple lesions, often in a parallel distribution and frequently following the lines of Blashko. Linear epidermal nevi may be small, measuring only a few centimeters or extensive, covering large portions of the body (Ichthyosis hystrix). It is the more extensive lesions that are most likely to be associated with underlying bony abnormalities and CNS symptoms.

Histology

Linear epidermal nevi resemble a benign papilloma histologically with considerable hyperkeratosis, papillomatosis and slight acanthosis (Fig. 3.5). When pseudo-horn cysts are present, they resemble a seborrheic keratosis. The surface resembles that of a nevus sebaceous, but they lack the dermal adnexal components. Some linear epidermal nevi may show focal histologic features of **epidermolytic hyperkeratosis**. Also, some lesions contain areas of **acantholytic dyskeratosis**.

Differential Diagnosis

Histologically, linear epidermal nevi resemble hyperkeratotic seborrheic keratosis, verruca vulgaris, hypertrophic actinic keratosis, and acanthosis nigricans.

Acrokeratosis Verruciformis of Hopf

Autosomal dominant, warty papules found on the dorsum of hands and feet and on lower legs. The papules are characterized by hyperkeratosis, hypergranulosis, mild papillomatosis with "church spire" elevations of the epidermis. This condition is sometimes associated with Darier's disease.

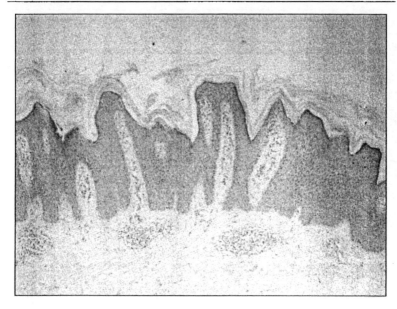

Fig.3.5. **Linear epidermal nevus**: Hyperkeratosis, papillomatosis, and irregular acanthosis.

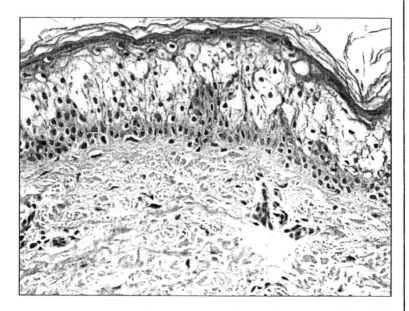

Fig.3.6. **Epidermolytic hyperkeratosis**: Most of the cells in the stratum of Malpighii are vacuolated and their cytoplasm appears shredded.

Table 3.2. Epidermal atrophy

Poikilodermatous dermatitis
Lichen sclerosus et atrophicus
Lentigo maligna
Striae distensae
Dego's disease (malignant atrophic papulosis)
Radiation dermatitis, late
Discoid lupus erythematosus (some cases)
Atrophic lichen planus (lichen planus actinicus)

Section 5. Epidermal Atrophy

Epidermal atrophy is clinically manifested by thin, translucent and sometimes scaly (cigarette paper) skin. Histologically, atrophy is manifested by a decrease in thickness of the stratum Malpighii with flattening and disappearance of the rete ridges. Since the epidermis varies in thickness in different regions, flattening of the epidermis is one of the best indicators of atrophy. Hyperkeratosis is not uncommonly seen with epidermal atrophy.

Degos' Disease (Malignant Atrophic Papulosis)

Clinical

Dego's disease presents with crops of yellow-red papules which umbilicate and gradually form on atrophic, porcelain-white center. The cutaneous lesions are asymptomatic, but gastrointestinal and central nervous system infarcts may occur. The etiology is unclear and the prognosis is generally poor.

Histology

Early lesions show swelling of the dermal collagen and an accumulation of mucin. The characteristic finding is a wedge-shaped ischemic infarct of the dermis, the apex being in the lower dermis. Within the infarct the collagen has a homogeneous appearance, and blood vessels and appendages are necrotic or absent. In late lesions mucin is present only at the margins of the infarct. The overlying epidermis is extremely thin, sometimes discontinued, with some degree of hyperkeratosis. The infarct is due to occlusion of an arteriole at the apex, but serial sections generally are required to find the affected arteriole. Occlusion occurs most commonly as the result of thrombosis. The wall of the vessel is often infiltrated by lymphocytes and histiocytes. Similar findings with thrombosis and inflammation of larger vessels occurs in the GI tract and the CNS with sometimes dismal results.

Striae Distensae

Clinical

Striae distensae usually occur on the abdomen, thighs, or buttocks often after pregnancy (striae gravidarum), excess corticosteroid (Cushing's) or systemic

Table 3.3. Disorders exhibiting epidermolytic hyperkeratosis

Bullous congenital ichthyosiform erythroderma
Linear epidermal nevus
Epidermolytic hyperkeratosis palmaris et plantaris
Verruca vulgaris
Isolated and disseminated epidermolytic acanthoma

corticosteriod therapy, heavy lifting, or in association with adolescence. They present as red linear bands of thin, atrophic skin which over time become purple, and then white.

Histology
The histology is characterized by a thin epidermis and dermis. Collagen bundles are thin and lie parallel to the skin surface in the upper dermis, resembling a scar.

Radiation Dermatitis, Late

Clinical
Late, or chronic, radiation dermatitis occurs months to years after exposure to radiation, most commonly after a short period of high dose radiation therapy. Clinically the skin becomes atrophic, hypo and hyperpigmented, and sclerotic. Telangiectasias are present and ulcerations may occur.

Histology
Histologically the epidermis is atrophic, focally hyperkeratotic, with occasional hydropic degeneration and dyskeratotic keratinocytes. The dermis appears homogeneous and sclerotic, with loss of skin appendages. Telangiectasiae and stellate, atypical "radiation fibroblasts" are frequently present.

Section 6. Granular Layer Alterations

Alterations in the granular layer occur in many disorders and can aide in their diagnoses. In normal skin, the thickness of the **stratum granulosum** is 1-3 cells thick, being proportional to the thickness of the stratum corneum. Characteristic changes in the thickness of this layer are found in lichen planus (wedge hypergranulosis), psoriasis (decreased), and ichthyosis vulgaris (decreased). Epidermolytic hyperkeratosis is a specific alteration of the granular layer characterized by exaggerated intra cytoplasmic vacuolization of the strati granulosum and spinosum, which acquire a reticulated appearance. The granular layer is markedly thickened with prominent keratohyalin granules. There is also hyperkeratosis (Fig. 3.6).

Bullous Congenital Ichthyosiform Erythroderma
This condition results from an autosomal dominant defect in keratins 1 and 10.

Clinical

Presents at birth with erythema and mild hyperkeratosis and frequently with denuded skin. Bullae form after mechanical trauma. Early in life, blistering is replaced by generalized hyperkeratosis with accentuation of flexural involvement.

Histology

Histologically, epidermolytic hyperkeratosis is present, and mitoses are numerous in the epidermis. Bullae, when present, are intraepidermal. The upper dermis contains a chronic inflammatory infiltrate.

Epidermolytic Keratosis Palmaris et Plantaris

Clinical

This abnormality is caused by an autosomal dominant defect in keratin 9 on chromosome 17Q. Presents in infancy as well demarcated palmoplantar hyperkeratosis. Clinically identical to keratosis palmaris et plantaris of Unna-Thost.

Histology

On histologic examination epidermolytic hyperkeratosis is present. The stratum corneum is extremely thickened.

Isolated and Disseminated Epidermolytic Acanthoma

Clinical

Clinically presents as a single warty papule or multiple flat brown papules. The differential diagnosis includes myrmecia warts and verruca vulgaris.

Histology

On microscopic examination, hyperkeratosis, papillomatosis as well as epidermolytic changes are present. The epidermolytic hyperkeratosis extends to, but does not include, the basal layer, similar to the histologic appearance of linear epidermal nevus with epidermolytic changes.

CHAPTER 4

Spongiosis, Exocytosis and Acantholysis

Ramón L. Sánchez and Sharon S. Raimer

Section 7. Spongiosis

Spongiosis is a process in which intercellular edema between the squamous cells of the epidermis causes an increase in the widths of the spaces between the cells. There is also intracellular edema manifested by intracytoplasmic vacuoles. When the intracellular edema is very prominent, reticular degeneration, characterized by bursting of cell walls with formation of vesicles, ensues. **Exocytosis** (i.e., migration to the epidermis) of lymphocytes is frequently seen in spongiosis (see section 8). **Spongiotic dermatitis** is characteristic of eczema and other inflammatory conditions of the skin.

Subacute Dermatitis

Clinical
Subacute dermatitis may occur with several clinical conditions, such as atopic dermatitis, contact dermatitis, stasis dermatitis, drug eruptions or generalized excoriated dermatitis. Edema and scaling of the skin are frequently present, and the patient frequently experiences pruritus.

Histology
Spongiosis is present and, if severe, reticular degeneration of epidermal cells may occur. Variable degrees of acanthosis are present, and in the stratum corneum there are areas of parakeratosis and crusting. A mononuclear cell infiltrate is present around superficial capillaries in the upper dermis (Fig. 4.1).

Acute Dermatitis

Clinical
In acute dermatitis, erythema of the skin with an edematous appearance is often present. Vesicles, bullae, and oozing may often be found. A typical example is contact dermatitis, both irritant or allergic (such as in **rhus dermatitis**).

Histology
Intraepidermally located vesicles and bullae are present in acute dermatitis. Spongiosis and intracellular edema is generally present in the epidermis surrounding the vesicles (Fig. 4.2). When large numbers of vesicles are present and the intracellular edema is pronounced, the histological picture is that of reticular degeneration of the

---

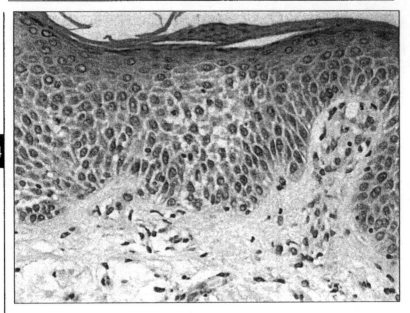

Fig 4.1. **Subacute dermatitis**: Focal parakeratosis, acanthosis, and spongiosis.

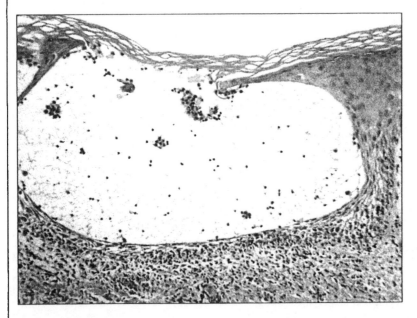

Fig 4.2. **Acute dermatitis**: Large intraepidermal spongiotic vesicle with occasional inflammatory cells.

Table 4.1. Nonvesicular spongiotic processes

Subacute dermatitis
Pityriasis rosae
Papular acrodermatitis of childhood (Gianotti-Crosti)

Table 4.2. Vesicular spongiotic processes

Acute dermatitis
Eczematous drug eruption

4

epidermis. Vesicles are separated from one another only by thin septae formed by the resisting walls of edematous epidermal cells and thus form a multilocular bullae. Vesicles and bullae, as well as the edematous portions of the epidermis, may be permeated by an inflammatory infiltrate composed mainly of mononuclear cells. In lesions that are several days old neutrophils are also present.

The stratum corneum may be parakeratotic and contain aggregates of coagulated plasma or crust. The dermis shows vascular dilatation, edema, and a mononuclear cell infiltrate. In the case of contact dermatitis, eosinophils, sometimes numerous, may be seen.

Oral White Sponge Nevus

Clinical
An autosomal dominantly inherited condition in which large areas, and sometimes the entire oral mucosa, have a thickened, folded, creamy white appearance. Occasionally the rectal mucosa, vagina, nasal mucosa or esophagus may be involved. Malignant degeneration is not known to occur.

Histology
Histologically there is hyperplasia of the epithelium with pronounced hydropic swelling of the epithelial cells. The focal swelling extends into the rede ridges but spares the basal layer. Nuclei appear smaller than normal. The surface shows parakeratosis, and only rarely are there small accumulations of keratohyalin granules.

Leukoedema of the Oral Mucosa

Clinical
A common condition which differs from white sponge nevus in that it is not inherited, develops in adulthood, is patchy rather than diffuse, and may have exacerbations and remissions.

Histology
In leukedema of the oral mucosa, as in white sponge nevus, the suprabasal epithelial cells show marked intracellular edema and nuclei that are smaller than normal (Fig. 4.3).

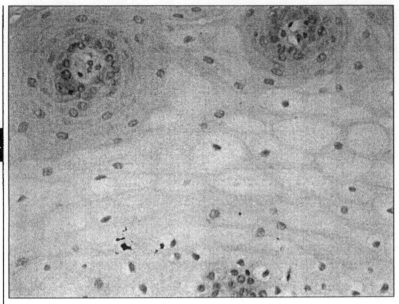

Fig 4.3. **Leukoedema of the oral mucosa**: Keratinocytes show clear, edematous cytoplasm.

Table 4.3. Leukoedema: Intracellular edema without spongiosis

Pachyonychia congenita (mucosa)
Oral white sponge nevus
Leukoedema of the oral mucosa
Oral focal epithelial hyperplasia
Oral hairy leukoplakia

Oral Focal Epithelial Hyperplasia

Clinical
A rare condition seen mainly in children. The lesions are limited to the oral mucosa, with the lower lip being most commonly affected. Soft white papules 2-4 mm in diameter are present. Most are discrete, but some are confluent.

Histology
Pathology is identical to that seen in oral epithelial nevus and in white sponge nevus.

Oral Hairy Leukoplakia

Clinical

This condition is most commonly found in individuals affected with human immunodeficiency virus. Clinically the lesions present as whitish plaques with a corrugated appearance on the lateral aspects of the tongue. Lesions may extend onto the dorsum or ventral surface of the tongue and less commonly are located on other mucosal surfaces.

Histology

Histologically oral hairy leukoplakia is characterized by acanthosis, hyperkeratosis, and upward projections of parakeratotic cells, resembling hairs. Koilocytes and vacuolated cells can be seen in the upper third of the prickle cell layer. Nuclear inclusion that contain Epstein-Barr virus can be found in routinely processed material. Surface colonization by *Candida* organisms and bacteria is common.

4

Section 8. Exocytosis/Epidermotropism

Exocytosis refers to the presence of mononuclear cells in the epidermis with spongiosis and often also associated microvesiculation occurring in various inflammatory dermatoses, especially subacute dermatoses. Exocytosis differs from epidermotropism which refers to the presence of mononuclear cells in the epidermis without spongiosis. This occurs in mycosis fungoides. Cells lie either singly, surrounded by a clear halo, or may be aggregated into **Pautrier's microabscesses**. The distinction between exocytosis and epidermotropism can be difficult, yet is crucial in establishing a correct diagnosis.

Pityriasis Lichenoides, Acuta and Chronica

Clinical

Pityriasis lichenoides occurs in two forms which differ in severity. Overlap between the two forms occur. The most severe and acute form of the disease, also known as **pityriasis lichenoides et varioliformis acuta** (PLEVA), also referred to as Mucha-Habermann disease, is generally characterized by crops of papules or papulo-vesicles that may develop hemorrhagic or necrotic centers. Individual lesions heal within a few weeks, generally with little or no scarring. The disease tends to be chronic,

---

Table 4.4. Exocytosis and epidermotropism

Exocytosis
        Spongiotic dermatitis
        Pityriasis rosea
        Pityriasis lichenoides, acuta and chronica
Epidermotropism
        Actinic reticuloid
        Lymphomatoid papulosis
        Mycosis fungoides

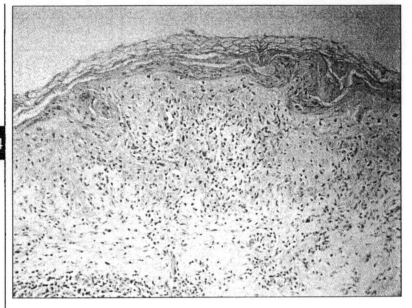

Fig 4.4. **PLEVA:** There is confluent parakeratosis, prominent exocytosis of lymphocytes, interface dermatitis, necrosis of keratinocytes, and extravasated RBCs.

recurring over a period of several months or sometimes even years. Occasionally
larger lesions, which develop into necrotic ulcers and result in scarring, develop.
Rarely a form of the disease that is characterized by ulcerative lesions which may
become confluent and are associated with fever and constitutional symptoms occurs.

The milder form of the disease known as **pityriasis lichenoides chronica** is characterized by recurrent crops of reddish-brown papules and scaly plaques which occur
mainly on the trunk. The disease tends to be chronic, persisting for months or even
years.

Histology

In the more severe form of the disease, known as PLEVA, the inflammatory
process has a wedge-like configuration and it presents as a mononuclear infiltrate
around capillaries, with the capillaries showing endothelial swelling and permeation
of their walls by the lymphocytes (this phenomena has been termed **lymphocytic
vasculitis**). This results in mild focal extravation of erythrocytes. There is an **interface dermatitis** often with marked exocytosis of lymphoid cells. A few erythrocytes
are usually seen within the epidermis, and this is a fairly characteristic feature of the
disease. Intercellular and intracellular edema may occur in the epidermis. Early lesions
show necrotic, dyskeratotic cells along the basal layer, while in more advanced lesions
the necrotic cells become more numerous and erosions or even ulcerations may be
present (Fig. 4.4). In general PLEVA differs from leukocytoclastic vasculitis by the
absence of neutrophils, nuclear dust, and fibrinoid deposits around capillaries,
however when ulceration occurs, these features may be present to some extent.

Pityriasis lichenoides chronica exhibits a perivascular infiltrate composed largely of mononuclear cells. The infiltrate does not invade the walls of the vessels and therefore extravasation of red cells are absent. When the infiltrate extends into the epidermis, one observes spongiosis. Individual necrotic keratinocytes are often seen, sometimes quite high in the epidermis rather than in the basal layer. A thickened, confluent parakeratotic horny layer is usually found.

Actinic Reticuloid

Clinical

Individuals with actinic reticuloid first show erythema and lichenified plaques in sun-exposed areas. The eruption gradually spreads to the covered areas of skin and may eventually result in a generalized erythroderma. The skin may show considerable thickening, resulting in deep furrows. Itching is severe.

Histology

A dense, band-like infiltrate is present in the upper part of the dermis and may extend into the lower dermis. The cells of the infiltrate are largely lymphoid cells and histiocytes with an admixture of eosinophils and plasma cells. Similarity to mycosis fungoides is often great because of the density of the infiltrate, and also because of the presence of cells with atypical, hyperchromatic nuclei resembling mycosis cells, which may invade the epidermis (**epidermotropism**) and form aggregates resembling Pautrier microabscesses.

Section 9. Acantholysis

Acantholysis occurs as a result of the loss of adhesion between epidermal cells which can result clinically in blisters and erosions, while histologically is manifested by intraepidermal clefts or vesicles formed by detached cells.

Pemphigus Vulgaris

Clinical

Pemphigus vulgaris is the most common form of pemphigus. Most patients present initially with lesions localized to the oral mucosa. Mucosal lesions may persist for several months before blisters appear on the skin. Cutaneous lesions start as flaccid blisters arising on normal or erythematous skin which rupture easily, resulting in large denuded areas that show little tendency to spontaneous healing. Involvement can be widespread. Nikolsky's sign is a typical finding in pemphigus vulgaris. The antigen in pemphigus vulgaris is desmoglein 3, located in the desmosomes of keratinocytes.

Histology

Loss of cohesion between epidermal cells, or acantholysis, leads to the formation of clefts, then bullae, in a predominantly suprabasal location. Basal cells, although separated from one another through the loss of intercellular bridges, remain attached

Table 4.5. Acantholytic processes

Pemphigus vulgaris
Pemphigus vegetans
Pemphigus foliaceous
Pemphigus erythematosus
Paraneoplastic pemphigus
Hailey-Hailey disease
Transient acantholytic dermatosis (Grover's)
Darier's disease
Warty dyskeratoma
Solitary/multiple acantholytic acanthoma
Actinic keratosis
Acantholytic squamous cell carcinoma
Pityriasis rubra pilaris (focal)
Acantholytic dermatosis of genito-crural region
Focal acantholytic dyskeratoma

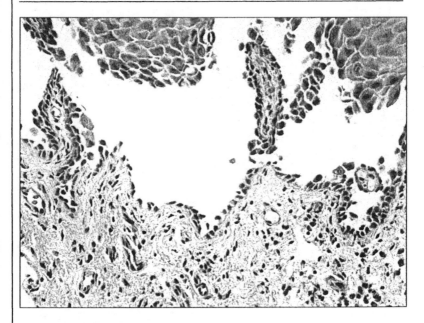

Fig 4.5. **Pemphigus vulgaris**: Acantholytic, suprabasal cleft with tombstone arrangement of the cells in the basal layer.

to the dermis like a row of tombstones (Fig. 4.5). There is generally little evidence of inflammation, although rarely eosinophils invade the epidermis before acantholysis becomes evident, and this has been referred to as **eosinophilic spongiosis**. The bullae in pemphigus vulgaris contains single, as well as clusters, of epidermal cells. These acantholytic cells appear rounded with large hyper chromatic nuclei and ho-

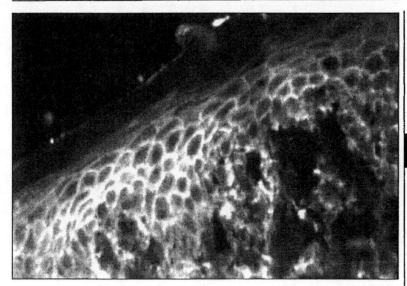

Fig 4.6. **Immunofluorescence pemphigus**: The characteristic intercellular positive fluorescence (chicken wire).

mogeneous cytoplasm. Frequently one observes on the floor of the bullae irregular upward growth of papillae that are lined by a single row of basal cells referred to as villi.

Immunofluorescence Testing

In direct immunofluorescence testing of the skin in patients with pemphigus vulgaris, unfixed frozen sections of perilesional skin are used. Fluorescein labeled antihuman, IgG, IgA, IgM, C3 and C4 are applied to various sections. With positive tests there is fluorescence of the intercellular spaces of the epidermis (**chicken wire**) (Fig. 4.6). IgG is almost invariably present, and in about one-half of patients there are deposits also of IgA, IgM, and/or C3. Immunofluorescence testing is positive in close to 100% of patients with pemphigus vulgaris. For indirect immunofluorescence testing, unfixed frozen sections of either guinea pig or monkey esophagus are used as substrate. The patient's serum and then fluorescein labeled antihuman IgG are applied at various dilutions to the sections. With positive tests there is fluorescence of the intercellular spaces of the esophagus mucosa. The highest dilution of the serum that still shows fluorescence indicates the antibody titer.

Pemphigus Vegetans

Clinical

There are two types of pemphigus vegetans, the Neumann type and the Hallopeau type. The Neumann type typically begins as pemphigus vulgaris, but instead of healing with normal skin, many of the denuded areas heal with verrucous vegetative

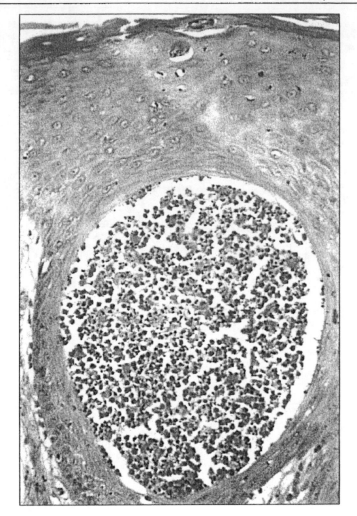

Fig 4.7. **Pemphigus vegetans**: Marked acanthosis and intraepidermal acantholytic pustule with neutrophils and eosinophils.

lesions, which are sometimes studded with small pustules. In the Hallopeau type of pemphigus vegetans pustules are the primary lesions. The occurrence is followed by the formation of gradually extending verrucous vegetative lesions, especially in intertriginous areas.

Histology

In the Neumann type of pemphigus vegetans, the early lesions show essentially the same histologic picture as pemphigus vulgaris, however the formation of villi and the downward proliferation of epithelial strands are much more pronounced than in pemphigus vulgaris. Verrucous vegetative lesions are characterized by papillomatosis and acanthosis. Though acantholysis is often no longer apparent, one not infrequently observes intraepidermal abscesses composed almost entirely of eosinophils. The abscesses are highly diagnostic of pemphigus vegetans (Fig. 4.7).

In the Hallopeau type of pemphigus vegetans early lesions show acantholysis with the formation of small clefts and cavities in a supra basal location. Cavities are filled with numerous eosinophils and acantholytic epidermal cells. A pronounced inflammatory infiltrate composed largely of eosinophils is present both in the epidermis, where it may present the histologic picture of eosinophilic spongiosis, and the dermis. Verrucous vegetative lesions show the same histologic picture as the Neumann type of pemphigus vegetans.

Pemphigus Foliaceous

Clinical

Pemphigus foliaceous is less common than pemphigus vulgaris, except in rural regions of Brazil where there is an endemic variant of this disease known as "Fogo Selvagem". Primary cutaneous lesions are superficial blisters that rupture easily leaving superficially denuded areas or patches that resemble dermatitis. Skin lesions are usually most prominent on the face, neck and trunk, however the disease may become generalized and present as large areas of erythematous dermatitis.

Histology

Changes in pemphigus foliaceous are characterized by areas of acantholysis in the upper part of the stratum Malpighii or right beneath it, with formation of a cleft in a superficial location. This cleft may develop into a bullae with acantholysis present at its floor, as well as at its roof (Fig. 4.8). More commonly, enlargement of the cleft leads to detachment of the uppermost epidermis without formation of a bullae. The dermis shows a moderate number of inflammatory cells among which eosinophils are often present. Immunofluorescence shows intercellular staining in the upper part of the epidermis, and indirect studies are generally positive.

Pemphigus Erythematosus

Clinical

The clinical appearance of pemphigus erythematosus is identical to that of pemphigus foliaceous except that the lesions usually remain primarily in sun exposed areas.

Histology

Histological appearance is also the same as pemphigus foliaceous. Direct immunofluorescence studies show in addition to the presence of intercellular antibodies in the upper part of the epidermis generally a "lupus band" in the uppermost dermis.

Fig 4.8. **Pemphigus foliaceous:** Acantholytic blister in the upper stratum of Malpighii. Acantholysis may be very sparse.

Paraneoplastic Pemphigus

Clinical

Paraneoplastic pemphigus (PNP) is a rare disease characterized by painful oral mucosal ulcerations and a variable skin eruption in association with an underlying malignancy, especially non-Hodgkin lymphoma. Cutaneous lesions may be vesicles, bullae, erosions, annular, polycyclic, or papulosquamous appearing on the trunk and extremities. They occasionally resemble targets, are confluent, or involve the palms and soles.

Histology

Basal layer acantholysis and intraepidermal clefts are present as are seen in pemphigus vulgaris. In addition vacuolar degeneration of the basal cell layer and the presence of necrotic keratinocytes (as in erythema multiforme) are characteristic findings of paraneoplastic pemphigus. Immunofluorescence of perilesional skin demonstrates the deposition of IgG with or without complement in the intercellular spaces of the epidermis. In addition granular or linear deposition of immunoreactants at the basement membrane site similar to that seen in bullous pemphigoid can be seen in some patients. Serum samples from patients with PNP contain circulating autoantibodies of the IgG class directed against the cell surface of stratified squamous epithelium. Indirect immunofluorescence studies should be performed on rodent bladder.

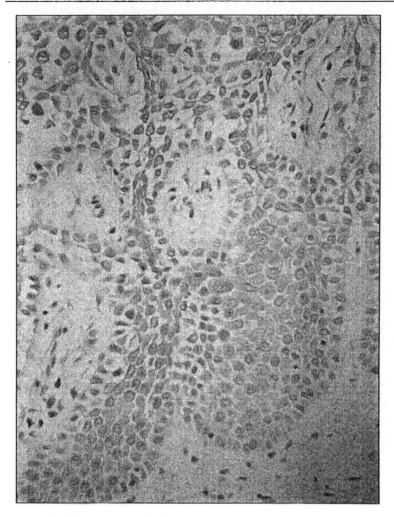

Fig 4.9. **Familial benign pemphigus (Hailey-Hailey):** There is separation and acantholysis of keratinocytes throughout the epidermis (dilapidated brick wall).

Familial Benign Pemphigus (Hailey-Hailey Disease)

Clinical

This condition is characterized by localized recurrent crops of vesicles occurring on an erythematous base generally in an annular configuration but occasionally circinate. The sites of predilection are intertriginous areas, particularly the axillae and groin. In intertriginous areas the vesicles are rapidly denuded, and the eruption may have the appearance of a chronic intertrigo.

Histology

Early lesions may show small superficial basal separation, so called lacuna, however fully developed lesions show large separations predominantly in a supra basilar location. Villi, which are elongated papillae lined by a single layer of basal cells, protrude upward into the bullae and, in some cases, strands of epidermal cells proliferate downward into the dermis. Acantholysis often affects large portions of the epidermis. Individual cells, as well as groups of cells, are seen in large numbers in the bullae cavity. The cells of the detached epidermis in many places show only slight separation from one another because of a few intact intercellular bridges which hold them loosely together. This typical feature gives the detached epidermis the appearance of a dilapidated brick wall (Fig. 4.9).

Transient Acantholytic Dermatosis (Grover's Disease)

Clinical

This entity is characterized clinically by the presence of pruritic, discrete papules or papulo-vesicles occurring mainly on the chest, back, and thighs. For the majority of patients, the condition is transient, lasting approximately two weeks to three months; however, in an occasional patient it is persistent and may be recurrent for over a period of years.

Histology

Small foci of acantholysis are present. Two or more histological patterns may be found in the same specimen. The most commonly seen pattern is that which resembles pemphigus vulgaris, showing supra basilar acantholysis (Fig. 4.10). A pattern resembling Darier's disease characterized by acantholytic dyskeratosis and the presence of corps ronds may also be present. In addition, there are patterns resembling Hailey-Hailey disease, with supra basilar clefts, and abundant acantholysis; a pattern resembling pemphigus foliaceus with acantholysis; and a spongiotic pattern in which acantholytic cells are present within spongiotic foci.

Darier's Disease (Keratosis Follicularis)

Clinical

This entity presents as keratotic papules occurring mainly in a seborrheic distribution on the scalp, the seborrheic areas of the face, and on the upper trunk. Occasionally the condition may occur extensively and may be found on the arms and hands and involve the majority of the trunk and legs. Atrophic lesions are occasionally present. Oral mucosa may be involved, in which case the mucosal surface has a pebbly appearance. The condition is inherited as an autosomal dominant trait.

Histology

Supra basilar acantholysis leads to the formation of supra basilar clefts or lacunae. The lacunae contain acantholytic cells which are devoid of intercellular bridges and show premature partial keratinization. A characteristic of this disease is a peculiar form of dyskeratosis which results in the formation of corps ronds and grains.

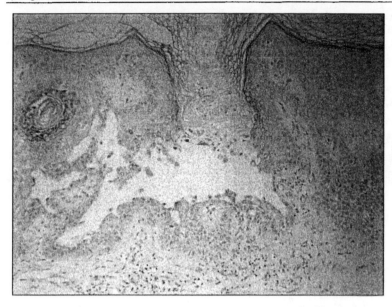

Fig 4.10. **Transient acantholytic dermatosis (Grover's)**: Acantholysis throughout the epidermis. Dyskeratotic cells similar to grains in Darier's Disease are often seen.

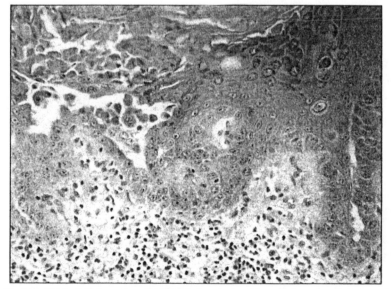

Fig 4.11. **Darier's disease**: Acantholytic dyskeratosis with multiple grains. A corp rond is seen on the right side of the photograph.

The corps ronds occur in the upper stratum Malphigii. The grains are found in the horny layer and with acantholytic cells in the lacunae. Corps ronds possess a central homogeneous basophilic, picnotic nucleus that is surrounded by a clear halo. Because of its size, the halo is conspicuous and the corps ronds stand out clearly. In contrast to the corps ronds, the grains are much less conspicuous and resemble parakeratotic cells that are somewhat larger. The nuclei of grains are elongated and often grain shaped, and are surrounded by homogeneous dyskeratotic material that usually stains basophilic, but may stain eosinophilic. In the lacunae are irregular upward proliferations of papillae lined with a single layer of basal cells known as villi (Fig. 4.11). The villi may be quite tortuous so that on histological examination, some of them appear in cross-section as rounded dermal structures lined by a solitary row of basal cells in the lacunae. In addition, there is papillomatosis, acanthosis, and hyperkeratosis. The dermis shows a chronic inflammatory infiltrate, and there may be downward proliferation of epidermal cells into the dermis. Lesions of the oral mucosa are analogous in appearance to those observed in the skin, although definite, well-formed corps ronds are generally absent.

Differential Diagnosis

Although acantholytic dyskeratosis in association with corps ronds is highly characteristic of Darier's disease, it occurs also in several other conditions. **Warty dyskeratoma** clinically is a solitary lesion with a deep central invagination. A similar histological pattern can occur in transient acantholytic dermatosis; however, clinically this entity consists of scattered discrete papules. **Focal acantholytic dyskeratoma** also presents clinically as a solitary papule and occasionally a few corps ronds may be seen in familial benign pemphigus.

Warty Dyskeratoma

Clinical

This entity is nearly always seen as a solitary lesion, most commonly on the scalp, face or neck. It has been reported to occur on other areas of the body, and has been observed on the oral mucosa. It most commonly presents as a papule or nodule with a keratotic, umbilicated center.

Histology

Histologically a large cup-shaped invagination occupies the center of the lesion and connects with the surface by a channel. The channel is filled with keratinous material. The large invagination contains numerous acantholytic dyskeratotic cells in its upper portion. The lower part of the invagination is occupied by numerous villi. At the base of the upward growing villi, irregular downward growth of epidermal strands are often seen as a double layer of basal cells separated by a narrow slit containing a few dyskeratotic acantholytic cells. Typical corps ronds can usually be seen in a thickened granular layer lining the channel at the entrance of the invagination.

CHAPTER 5

Vesicles and Bullae

Ramón L. Sánchez and Sharon S. Raimer

Section 10. Vesicles and Bullae

A blister is fluid filled space in the epidermis or immediately below the epidermis. Depending on the size, blisters are classified as vesicles when less than 1cm in diameter or bulla, larger than 1cm. In addition to fluid, blisters often contain inflammatory cells and/or epidermal cells.

The vesicular dermatoses are classified regarding the location of the vesicle in relation to the epidermis, as well as the mechanism of blister formation. Regarding location, vesicles and bulla are classified as subcorneal, intraepidermal, or subepidermal. The intraepidermal vesicles are sometimes specified as suprabasal, when only the basal layer remains on the floor of the blister (characteristic of pemphigus vulgaris). Likewise, the subepidermal vesicles can be specifically classified as junctional, when the split takes place at the basal membrane, usually in the lamina lucida (bullous pemphigoid), and intradermal when all of the basal membrane is on the roof of the blister (epidermolysis bullosa acquisita). The blisters in epidermolysis bullosa simplex appear histologically as subepidermal while technically they are intraepidermal being formed by destruction of the cells of the basal layer.

Intraepidermal Vesicles

Spongiotic Intraepidermal Vesicles, (see section 7)

Acantholytic Blisters, (see section 9)

Ballooning Degeneration Blisters

Ballooning degeneration implies marked swelling of the cytoplasm of the keratinocytes without associated spongiosis. The cells become rounded, lighter in color and detached from adjacent cells, becoming acantholytic. It is seen in viral processes, particularly Herpes virus. Related to ballooning is reticular degeneration, characterized by marked edema of the keratinocytes that finally burst, forming vesicles in which thin strands of cytoplasm of the keratinocytes still remain.

Herpes Simplex, Varicella/Zoster

Clinical

Herpes simplex is one of the most common viral infections of the skin, and is caused by HSV-1 and HSV-2 (**Herpes simplex hominis**). The primary infection

---

Table 5.1. Subcorneal blisters with neutrophils (see section 19)

Dermatophyte/*Candida* infections
Subcorneal pustular dermatosis (Sneddon and Wilkinson)
Pustular drug eruption
Impetigo
Pemphigus erythematosus/foliaceous
Necrolytic migratory erythema
Transient neonatal pustular melanosis

Table 5.2. Subepidermal blisters

Bullous pemphigoid
Cicatricial pemphigoid
Dermatitis herpetiformis
Linear IgA dermatosis
Porphyria cutanea tarda/coproporphyria
Epidermolysis bullosa acquisita
Bullous diabeticorum
Amyloid bulla
Erythema multiforme
Polymorphous light eruption (vesicular)
Pemphigoid gestationes
Blisters arising in scars
Toxic epidermal necrolysis
Bullous lupus erythematosus
Bullous lichen planus
Coma/Pressure induced bullae

---

with HSV-1 usually occurs during childhood, is almost universal and often involves the mouth and throat (gingivitis, pharyngitis). Infection with HSV-2 is often sexually transmitted, although it can be acquired also by fomites. Varicella/Herpes Zoster virus, also in the family of Herpes viruses, presents initially as chicken pox.

The severity of symptoms of Herpes simplex infection is variable, frequently more severe in primary than in recurrent infection. The lesions are characterized by grouped vesicles on an erythematous base, often painful. With time the vesicles may become pustules and dry up with formation of crust. Associated symptoms such as urethritis and lymphadenopathy may occur. The course is of about two weeks duration. Recurrent lesions are usually preceded by a burning sensation, and in the case of **Herpes labialis** they may be triggered by external factors such as sun burn or by fever. Herpes zoster lesions are also characterized by grouped vesicles on an erythematous base, usually involving more extensive areas with a dermatomal distribution. Intense pain frequently precedes the lesions and postherpetic neuralgia may follow resolution.

Histology

The microscopic findings are similar for both herpes simplex and varicella/zoster. In a fully developed lesion there are acantholytic intraepidermal vesicles, the result

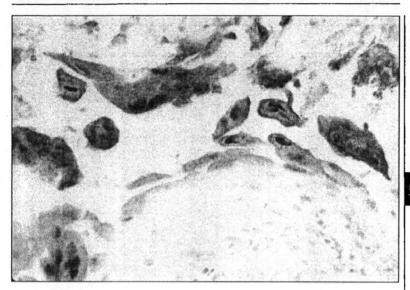

Fig 5.1. **Herpes**: Immunoperoxidase stain for Herpes virus shows positive results (brown stain) in multinucleated giant cells.

of ballooning and reticular degeneration. Often the blisters are broken by the time the biopsy is obtained, with formation of ulcers marginated by diagnostic cells. Individual cells in the adjacent epidermis may appear swollen with clear cytoplasm and ground-glass homogenization of the nuclei, which acquires a bluish gray color with margination of the chromatin. These intranuclear viral inclusions are diagnostic (Cawdry type A).

The characteristic histologic findings are found in the blisters. Acantholytic multinucleated cells are seen displaying large size and molding nuclei (shape accommodates to neighboring nuclei) and ground-glass intranuclear viral inclusions (Fig. 5.1). Other cells appear as large tear-shaped dyskeratotic cells with shrunken, hyperchromatic nuclei and brightly eosinophilic cytoplasm, resembling the grains in Darier's disease. The multinucleated giant cells can usually be obtained by scraping the base of blisters of Herpes/Varicella and then smear the material on a glass slide and staining with Giemsa or Toluidine blue (Tzanck smear). The Tzanck smear is an easy, rapid cytologic test which can be performed in the clinic. The dermis underneath the blisters characteristically shows a marked lymphocytic and neutrophilic infiltrate, superficial and deep. Some degree of necrosis is often seen. Vasculitis, sometimes with viral inclusions in endothelial cells may be present. Occasionally the hair follicles are involved (**Herpes folliculitis**) and the characteristic findings may be seen only in the epithelium of the follicle, surrounded by extreme dermal inflammation.

Hydroa Vacciniforme

Clinical

It is characterized by the presence of erythema and discrete vesicles in sun exposed areas, particularly the face following sun exposure. The vesicles result in crusted lesions and eventually in scars. This is a rare condition seen most commonly in children.

Histology

The vesicles are intraepidermal and are formed by reticular degeneration, which eventually may evolve into full epidermal necrosis. The dermis shows acute and chronic inflammation, superficially and deep, with occasional eosinophils.

Subepidermal Blisters

Bullous Pemphigoid

Clinical

Chronic vesicular or bullous dermatosis is seen mainly in elderly patients. Clinically there are two types or stages of lesions: **erythematous macules** to **urticarial plaques** which eventually evolve into blisters, and the well established vesicular stage with tense, often intact **bulla** that can measure several centimeters in diameter and contain a clear fluid and may be present on an erythematous or nonerythematous base. The bulla are typically located on the abdomen and the flexor surface of extremities, although they can be seen elsewhere. Mouth involvement is seen in a small number of patients. The disease runs a protracted course. Occasionally bullous pemphigoid has been associated with autoimmune diseases and with several medications, including sulfasalazine, ibuprofen, phenacetin, furosemide, penicillin, and others.

Histology

The characteristic finding in bullous pemphigoid is the presence of a subepidermal blister which contains clear fluid as well as a number of inflammatory cells including neutrophils, some lymphocytes and usually a conspicuous number of eosinophils (Fig. 5.2). The separation between the epidermis and the dermis is generally sharp. The dermis underneath the bulla displays an inflammatory infiltrate similar to that described inside the blister. Occasionally in old lesions there is re-epitheliation of the base of the bulla, which starts in pre-existing hair follicles or from the adjacent normal epidermis, and the bulla appears intraepidermal. However, the roof in re-epitheliazed bullae shows a complete epidermis with preservation of the rete ridges. Secondary changes such as infection following rupture of the bulla can be seen and can distort the histologic appearance.

Early erythematous lesions of bullous pemphigoid may not show a blister, and the histologic findings may be limited to **eosinophilic spongiosis** which is manifested by the presence of eosinophils in the stratum malphighii, either individually or in groups.

Table 5.3. Eosinophilic spongiosis

Bullous pemphigoid
Pemphigus vulgaris
Pemphigus vegetans
Insect bites
Contact dermatitis
Erythema toxicum neonatorum
Incontinentia pigmenti

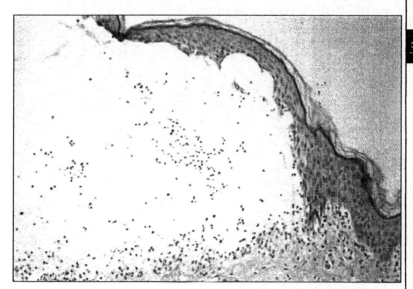

Fig 5.2. **Bullous pemphigoid**: A subepidermal blister with inflammatory cells including abundant eosinophils.

Immunofluorescence

Direct immunofluorescence on perilesional skin of bullous pemphigoid shows linear deposition of IgG and C3 along the basal membrane (Fig. 5.3). IgM and IgA sometimes are also positive. On salt split skin, the immunoreactants appear on the roof of the bulla. The bullous pemphigoid antigen appears to be associated with the lamina lucida and the hemidesmosomes of the basal keratinocytes.

Electron microscopy

The split takes place along the lamina lucida with destruction of the hemidesmosomes.

Differential Diagnosis

The bullae in **dermatitis herpetiformis** show abundant neutrophils but few if any eosinophils. There are also microabscesses in the dermal papillae. **Epidermolysis**

Table 5.4. Papillary neutrophilic microabscesses

Dermatitis herpetiformis
Cicatricial pemphigoid
Bullous lupus erythematosus
Linear IgA bullous dermatosis
Pemphigoid gestationes (with eosinophils)

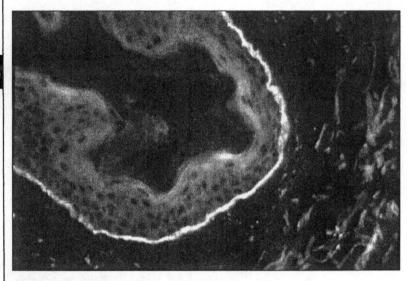

Fig 5.3. **Immunofluorescence of bullous pemphigoid**: Linear IgG deposition along the basement membrane.

bullosa acquisita (EBA) is usually less cellular, but often the differential diagnosis rests on immunoflourescence (in salt split preparations the immunoreactants are in the dermal side in EBA), or electron microscopy (decrease or absence of anchoring fibrils).

Cicatricial Pemphigoid

Clinical

It is a chronic bullous disease which characteristically involves mucous membranes particularly the mouth and conjunctiva, and resolves with atrophic scars which may result in blindness and respiratory compromise. In the Brunsting-Perry type of cicatricial pemphigoid there is less severe mucous membrane involvement and recurrent blistering occur outside of the mucous membranes. The areas more commonly involved are the face, scalp and upper chest. Cicatricial pemphigoid has been associated with autoantibodies to the bullous pemphigoid minor antigen (BPAg2) and in some cases with autoantibodies to epilegrin.

Histology

There is a subepidermal blister and the split characteristically extends into the hair follicles. The adjacent dermis often shows fibrosis with marked acute and chronic inflammation, in which eosinophils are present.

Direct immunofluorescence shows linear deposits of IgG and C3 along the basal membrane.

Electron microscopy shows the split along the lamina lucida in the basal membrane.

Dermatitis Herpetiformis

Clinical

Vesicular dermatosis characterized by the presence of papulovesicles on elbows, buttocks, and other extensor surfaces with a symmetrical distribution. The condition is extremely pruritic which results in excoriations. Early vesicles are preferred for histologic diagnosis. A majority of patients have an associated gluten sensitive enteropathy, and a gluten free diet usually reverses the skin condition. Most patients with dermatitis herpetiformis have IgA class endomysial antibodies. They are probably formed in the intestine and deposited in the dermal papillae producing the disease.

Histology

A well developed blister shows a subepidermal split and a large number of neutrophils, both inside the vesicle and in the surrounding dermis. Characteristically, the adjacent dermal papillae show neutrophilic microabscesses. Biopsy from an early lesion may show only the papillary microabscesses without vesicle formation. Eosinophils may be present although they are usually not as numerous as in bullous pemphigoid (Fig. 5.4).

Immunofluorescence

Granular deposits of IgA are seen in the dermal papillae of normal skin. IgM as well as C3 also may be found. For immunofluorescence studies, the biopsies are preferably obtained from clinically uninvolved skin.

Linear IgA Bullous Dermatosis

Clinical

This condition is seen in two different settings: as bullous dermatosis of childhood, and as adult linear IgA bullous dermatosis. In the first instance, the disease presents in the first decade of life as tense bulla on normal or erythematous skin. The bulla has a predilection for the face, groin, upper and lower extremities. In the adult linear IgA bullous dermatosis there is development of tense, pruritic bullae particularly on extremities and trunk. Mucosal involvement may occur. These patients lack gluten sensitivity enteropathy, contrary to what is seen in dermatitis herpetiformis. They also lack IgA antiendomysial antibodies. Adult linear IgA dermatosis is sometimes associated with drugs including Captopril, Bactrim, Lithium, Amiodarone, Phenytoin, Vancomycin, and some of the nonsteroidal anti-inflammatory drugs. In

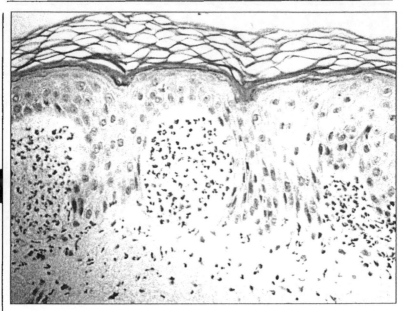

Fig 5.4. **Dermatitis herpetiformis**: There are neutrophilic microabscesses in the dermal papillae, with early separation and blister formation.

those instances, the condition resolves several weeks after discontinuation of the offending drug, while in the usual adult case the course is protracted.

Histology

The histologic findings are indistinguishable from those of dermatitis herpetiformis, including the presence of papillary neutrophilic microabscesses with fibrin deposition, focal kariorrhexis, and presence of subepidermal bullae with neutrophils.

Immunofluorescence

Direct immunofluorescence shows the presence of linear IgA deposition along the basal membrane on nonlesional skin. IgG, IgM and C3 may be also present. The IgA molecule is only IgA_1 as opposed to dermatitis herpetiformis in which both IgA_1 and IgA_2 are present.

Porphyria Cutanea Tarda (PCT)

Clinical

This condition is the most commonly seen form of porphyrias. It is characterized by reduced activity of **uroporphyrinogen decarboxylase** (URO-D) which is the enzyme that catabolizes the transformation of uroporphyrinogen into coproporphyrinogen. The enzyme is usually produced in the liver. PCT is seen as a familiar form, a sporadicform, or a toxic form. The familiar form is seen earlier in life. It is transmitted as an autosomal dominant trait and there is a deficiency of URO-D in the liver as well as in the erythrocytes. In the sporadic form with the onset in mid-life, most instances are associated with alcohol abuse and liver damage, but other triggering factors are estrogen therapy, oral contraceptives, diabetes mellitus, hepatitis C infection, Wilson's disease, hepatocellular carcinoma, lymphoma, and HIV infection. The toxic form results from exposure to polychlorinated aromatic hydrocarbons. Clinically PCT presents with vesicles and bulla on the dorsal aspect of hands and forearms which eventually rupture forming ulcers and scars. There is also hypertrichosis, typically on malar areas. Other manifestations may include patches of alopecia and sclerodermoid changes in the skin.

Histology

The typical lesion of PCT is a noninflammatory epidermal blister containing only clear fluid. The floor of the bulla shows festooning as a result of porphyrin deposition in the basal and sub-basal membrane zone (Fig. 5.5). Other findings are reduplication of basal membrane and deposition of an eosinophilic material in the papillary dermis and around superficial capillaries. The presence of eosinophilic material in the papillary dermis is more prominent in other types of porphyria such as erythropoietic protoporphyria than in PCT, and it may resemble lichen amyloid or colloid milium. Immunofluorescence may show deposition of IgG and sometimes IgM and C3 around the upper dermal blood vessels.

The histologic findings in PCT are indistinguishable from those seen in **coproporphyria**, which is an autosomal dominant variant of acute porphyria resulting from a deficiency of **coproporphyrinogen oxidase**.

Pseudoporphyria refers to a bullous dermatosis on sun exposed areas indistinguishable histologically from PCT, although it lacks the elevated levels of porphyrinogen in the urine and in feces. Pseudoporphyria has been associated with exposure to several drugs including tetracyclines, sulphonamides, isotretinoin, nalidixic acid and naproxen among others. It has also being seen in patients with chronic renal failure undergoing dialysis, and sometimes with long use of tanning beds.

Epidermolysis Bullosa (EB)

Clinical

Epidermolysis bullosa constitutes a series of blistering disorders that are generally transmitted with an autosomal dominant or recessive type of inheritance. The inherited forms of epidermolysis bullosa are roughly classified in three groups, depending on the level at which the blisters form: they may be produced in the basal layer of the epidermis (**epidermolytic** EB), at the level of the basal membrane, generally lamina lucida (junctional EB), or at the level of the dermis (dystrophic EB). The classic presentation of the epidermolytic form is EB simplex which often has a mild clinical presentation, although more severe generalized forms exist. Of the junctional EB forms, there is EB gravis, Hurlitz type, junctional EB associated with pyloric stenosis, and EB atrophicans mitis which has a better prognosis. The inherited

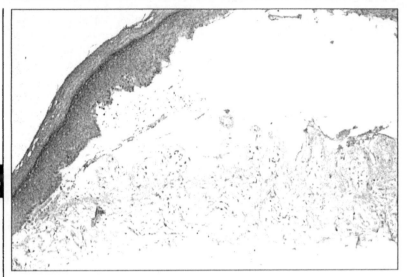

Fig 5.5. **Porphyria cutanea tarda**: Subepidermal, non-inflammatory blister. The floor of the blister shows festooning.

dystrophic forms of EB are characterized by extensive blistering and erosions which heal with scarring. Permanent nail loss is characteristic and frequently oral and gastrointestinal lesions such as esophageal webs and strictures are present. In severe forms of the disease webbing of the fingers and toes occurs. Squamous cell carcinomas frequently develop in the chronic ulcerations of dystrophic EB.

Epidermolysis Bullosa Acquisita (EBA)

Clinical

This is an acquired form of dermolytic epidermolysis bullosa with usual onset in adulthood, characterized by the presence of bullae in areas of minor trauma. Occasionally, patients may present with extensive bullae on the trunk and extremities that resemble bullous pemphigoid or cicatricial pemphigoid. The disease is produced by formation of autoantibodies to the epidermolysis bullosa acquisita (EBA) antigen, which binds to the collagen VII of anchoring fibers.

Histology

EBA is characterized by subepidermal blisters that are inflammatory. Papillary neutrophilic microabscesses can be seen, and the histologic findings are difficult to distinguish from those seen in bullous pemphigoid or occasionally in dermatitis herpetiformis (Fig. 5.6). PAS stain often will show the basal membrane attached to the roof of the blister. Likewise, laminin and type IV collagen can be identified by immunoperoxidase in the roof of the blister. Direct immunofluorescence shows linear deposition of IgG and complement in the basal membrane zone. Circulating IgG

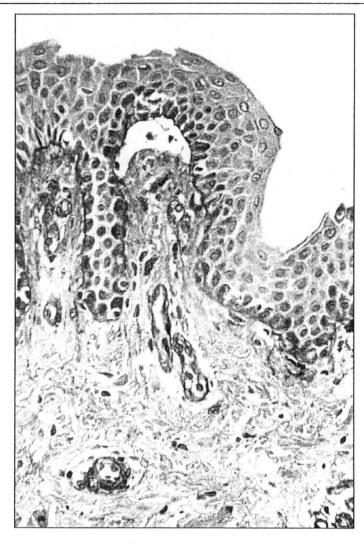

Fig 5.6. **Epidermolysis bullosa acquisita (EBA)**: An early subepidermal blister, stained for type IV collagen which appears in the roof of the blister.

antibodies are present in about 50% of cases. Using salt split skin the antibodies fix to the floor of the blister as opposed to bullous pemphigoid in which there is binding to the roof of the blister. Immunoelectronmicroscopy has shown the antibodies to be deposited below the lamina densa, sometimes covering the anchoring fibrils.

Pemphigoid (Herpes) Gestationes

Clinical

It is a vesiculo-bullous disease that is usually seen in the last trimester of pregnancy, although it can be seen in the second trimester or may develop postpartum. Sometimes it is associated with hydatidiform moles. Very pruritic papules and urticarial plaques which evolve to form vesicles and bullae develop initially in a periumbilical distribution with spread to the rest of the trunk and extremities.

Histology

The dermis shows a mixed infiltrate including neutrophils and eosinophils. There is usually formation of subepidermal vesicles and bullae which sometimes may appear to be located intraepidermally due to the orientation of the tissue. The bullae contain eosinophils, lymphocytes and histiocytes (Fig. 5.7).

Direct immunofluorescence shows deposition of C3 and sometimes IgG in a linear pattern along the basal membrane. The immunoglobulins are located in the lamina lucida and on anchoring filaments similar to bullous pemphigoid. A circulating autoantibody of the IgG class, the so called pemphigoid gestationes factor, can be found in the majority of patients with pemphigoid gestationes.

Other subepidermal bullous processes include:

Diabetic Bullae

Manifests as tense bulla on the lower extremities of patients with diabetes mellitus without renal failure. Histologically the bullae are noninflammatory, subepidermal and sometimes intraepidermal.

Amyloid Bullae

Noninflammatory, subepidermal bullae may be seen sometimes in amyloidosis. The surrounding dermis shows deposition of amyloid.

Erythema Multiforme (Bullous)

Shows subepidermal bullae formed by pronounced edema of the papillary dermis, which also contains fibrin and inflammatory cells. Other changes of erythema multiforme are seen around and within the bulla.

Toxic Epidermal Necrolysis

Shows full thickness necrosis of the epidermis which is detached from the dermis, with formation of a bullae or an empty space. Noninflammatory.

Bullous Lupus Erythematosus

Characterized by the presence of neutrophils and nuclear dust at the dermal-epidermal junction with separation of the epidermis from the dermis. Otherwise changes of systemic lupus erythematosus are seen particularly around hair follicles and other adnexae.

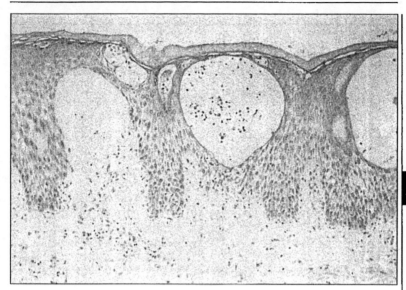

Fig 5.7. **Herpes gestationes**: Subepidermal blister with lymphocytes and eosinophils. The epidermis appears vesiculated, probably due to the cut orientation.

Lichen Planus

Sometimes exhibits cleavage between the epidermis and the dermis, the so-called Max-Joseph spaces, in an otherwise typical lichen planus picture. Full separation of the epidermis may be seen.

Pressure and Coma Induced Bullae

Are usually subepidermal although the epidermis may have spongiosis which may result also in intraepidermal vesicles. When the bulla is associated with barbiturate intake there is necrosis of ecrine coils and ducts which is diagnostic for this entity.

CHAPTER 6

Interface, Poikilodermatous Dermatitis and Necrosis of Keratinocytes

Ramón L. Sánchez and Sharon S. Raimer

Section 11. Interface Dermatitis

Interface dermatitis is characterized by the presence of inflammatory cells at the dermo-epidermal junction, associated with vacuolar degeneration of the cytoplasm of the basal cells, so-called hydropic degeneration. When this phenomenon is very prominent it may result in separation of the epidermis from the dermis. The inflammatory cells are usually lymphocytes and sprinkle from the capillaries of the papillary dermis. Interface dermatitis should be distinguished from **lichenoid dermatitis**, in which there is a lichenoid infiltrate in addition to the interface changes.

Drug Eruption

Clinical

Multiple clinical presentations including morbilliform, eczematous, photodistributed, lichenoid, and fixed drug eruptions may occur. The latter are characterized by large bulla that resolve with hyperpigmented macules, always in the same location after every exposure to the causative drug. Drugs may also cause erythema multiforme, urticaria, and toxic epidermal necrolysis. Bullous pemphigoid and linear IgA dermatosis occasionally are precipitated by drugs. Leukocytoclastic vasculitis, acneiform reactions, and hyperpigmentation may be manifestations of drug reactions. Among the latter, tetracycline, doxycycline, minocin, antimalarials and amiodarone may be associated with hyperpigmentation of the dermis. The pigment in those cases usually stains positive for melanin (fontana stain), although occasionally it also gives a positive reaction for iron (Perl's stain).

Histology

The typical histologic findings in exanthematous, morbilliform reactions are interface dermatitis with edema of the papillary dermis and a mixed infiltrate of lymphocytes, variable number of eosinophils, and occasional neutrophils or plasma cells (Fig. 6.1). Not all biopsies show the full histologic picture described. Often biopsies of drug reactions will show only some of the findings, particularly a mixed infiltrate with some eosinophils and mild edema of the papillary dermis. In those instances, the diagnosis of drug reaction is assumed based on the clinico-pathologic correlation. The dermatopathologist should convey to the clinician that the findings although not diagnostic, are consistent with a drug reaction.

Dermatopathology, edited by Ramón L. Sánchez and Sharon S. Raimer. ©2001 Landes Bioscience.

Table 6.1. Interface dermatitis

Drug eruption
Erythema multiforme
Lichen planus
Lupus erythematosus
Pityriasis lichenoides et varioliformis acuta (PLEVA)
Parapsoriasis
Erythema dyschromicum perstans
Graft vs. host disease
Lichen sclerosus et atrophicus
Poikilodermatous dermatitis

Fixed Drug Reactions

When the biopsy is obtained from an active bullous lesion, it shows marked edema of the papillary dermis with bulla formation, interface dermatitis with marked swelling of the cytoplasm of the basal cells, and a lymphocytic infiltrate with eosinophils. There may be occasional necrotic keratinocytes, and usually there is prominent pigment incontinence (Fig. 6.2). Old lesions of fixed drug show mainly pigment incontinence in melanophages of the papillary dermis with a mild lymphocytic infiltrate.

Differential diagnosis of drug eruptions include **erythema multiforme**, established mainly by the presence of **necrotic keratinocytes** in the basal layer or above in erythema multiforme, and its absence in drug eruption. **Viral exanthem** can show similar histologic (and clinical) changes as in drug reactions, although there is often an absence of eosinophils. Some cases of parapsoriasis show mainly interface dermatitis with few changes in the epidermis, although parakeratosis is usually present.

Graft vs Host Disease (GVHD)

Clinical

It is mainly seen following bone marrow transplantation for treatment of leukemia or lymphoma. It is occasionally seen in infants after maternal-fetal transfusion in-utero. GVHD is classified in two stages, acute and chronic. Acute GVHD refers to the first eighty days after bone marrow transplant and it is characterized by vomiting, diarrhea, hepatic manifestations and an erythematous macular rash. Chronic GVHD is manifested by an early lichenoid phase which resembles lichen planus, followed by a poikilodermatous phase which usually concludes with a sclerodermoid phase. The pathogenesis of GVHD is attributed to the interaction of the donor lymphocytes with the histocompatibility markers of the recipient.

Histology

In the early stage there is an interface dermatitis in which lymphocytes invade the lower layers of the epidermis resulting in individual necrotic keratinocytes. Sometimes the lymphocytes are seen in close approximation or surround necrotic keratinocytes, a phenomena referred to as satellite cell necrosis (lymphocyte-associated apoptosis). In the chronic phase the lichenoid lesions resemble lichen planus

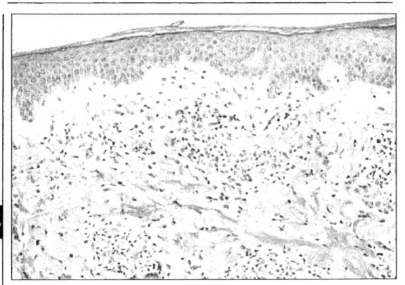

Fig 6.1. **Drug reaction**: There is edema of the papillary dermis and a perivascular lymphocytic infiltrate with eosinophils.

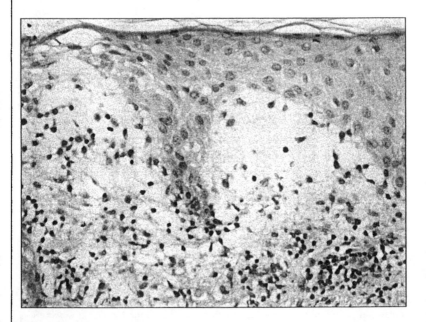

Fig 6.2. **Fixed drug reaction**: Prominent upper dermal edema with vesicle formation and a mixed infiltrate among which melanophages are seen.

while in the late sclerodermoid phase there is thickening of the collagen bundles which may result in atrophy of skin appendages.

Section 12. Poikilodermatous Dermatitis

Poikilodermatous dermatitis is characterized clinically by a thin, atrophic epidermis which is somewhat translucent and therefore shows the underlying vascular reticulated pattern. Since there is usually hyperkeratosis it has also been described as having the appearance of cigarette paper. Histologically, poikilodermatous dermatitis shows hyperkeratosis with follicular plugging, atrophy of the epidermis with absence of rete ridges, vacuolar interface dermatitis with a sprinkling of lymphocytes at the dermo-epidermal junction, an occasional lichenoid infiltrate, dilated capillaries and/or **telangiectasia**, and pigment incontinence.

Poikilodermatous dermatitis is seen in three genodermatoses, and as a manifestation of three systemic conditions. The genodermatoses are:

Poikiloderma congenitale (Rothmund-Thompson syndrome) is an autosomal recessive disorder characterized by a reticular erythematous eruption within the first year of life. Areas of hyperpigmentation, warty keratoses, short stature, cataracts, hypogonadism, mental retardation, photosensitivity, and the occasional development of skin cancer are characteristic of the syndrome.

Congenital telangiectatic erythema (Bloom's syndrome) is an autosomal recessive disorder characterized by facial sun-sensitive and telangiectatic rash, stunted growth, tendency to respiratory and gastrointestinal infections, chromosomal abnormalities and a variety of congenital malformations.

Dyskeratosis congenita is characterized by a reticular hyperpigmentation, nail dystrophy and leukokeratosis of mucus membranes. The development of squamous cell carcinomas is likely.

Among systemic acquired conditions manifested as poikilodermatous dermatitis are large plaque parapsoriasis or early mycosis fungoides, systemic lupus erythematosus, and dermatomyositis.

Large Plaque Parapsoriasis/Early Mycosis Fungoides

Clinical

Several terms have been used to describe this condition including **poikiloderma vasculare atrophicans** and large plaque parapsoriasis which is clinically characterized by thin atrophic skin with reticulated, mottled hyperpigmentation.

Histology

Histologically shows hyperkeratosis, epidermal atrophy, hydropic degeneration of basal layer, pigment incontinence, telangiectasia of superficial capillaries, and a band-like infiltrate which not only reaches the basal layer but also penetrates focally into the epidermis. The lymphocytes may show cerebriform nuclei and exhibit a clear halo when located in the epidermis. Early, patch stage mycosis fungoides may

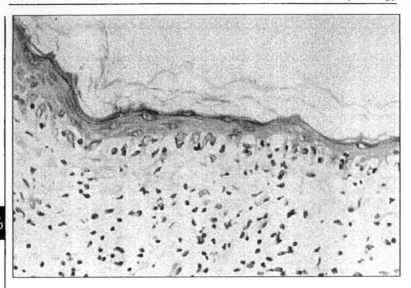

Fig 6.3. **Poikilodermatous dermatitis**: Epidermal atrophy, vacuolar interface dermatitis (hydropic degeneration), lymphocytic infiltrate and telangiectasiae.

show similar histologic findings although there is more prominent epidermotropism and occasional **Pautrier microabscesses**.

Clinical

A related condition is **hypopigmented mycosis fungoides**, characterized clinically by hypopigmented macules on the trunk and extremities without erythema.

Histology

Histologically the upper dermis shows a mild band-like lymphocytic infiltrate with an almost continuous lentiginous-like single cell row of lymphocytes along the basal layer of the epidermis. The lymphocytes characteristically are surrounded by a clear halo and are probably responsible for the hypopigmentation. These findings have been interpreted as the earliest diagnostic features of mycosis fungoides (Fig. 6.3).

Systemic Lupus Erythematosus (SLE)

Clinical

The skin manifestations of SLE include malar erythema manifested by slightly indurated erythematous patches on the face in a butterfly configuration, as well as erythematous patches and annular plaques on trunk and upper extremities. Occasionally SLE shows urticarial, purpuric, ulcerated or bullous lesions. Vasculitis is also a manifestation of SLE. Discoid lesions can be seen occasionally. Late manifestations of SLE include poikiloderma vasculare atrophicans.

Table 6.2. Criteria for the classification of SLE

Malar rash
Discoid rash
Photosensitivity
Oral ulcers
Arthritis
Serositis (pleurisy or pericarditis)
Renal disorder
Neurologic disorder (seizures or psychosis)
Hematologic disorder (anemia, leukopenia, thrombocytopenia)
Immunologic disorder (positive LE, anti-DNA abnormal titer, antibody to Sm
nuclear antigen, positive test for syphilis)
Abnormal ANA titer

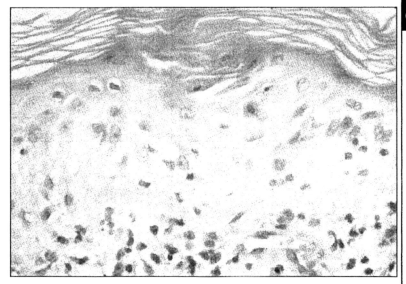

Fig 6.4. Systemic lupus erythematosus: The histologic findings are those of poikilodermatous dermatitis with eosinophilic, thick subbasal membrane zone.

In general a diagnosis of SLE can be established when there are three of the following four signs: cutaneous eruption consistent with LE, renal involvement, serositis, and joint involvement.

Histology

SLE is usually characterized by a **poikilodermatous dermatitis** with a sparse lymphocytic infiltrate. There is thickening of the basal membrane zone by deposition of an eosinophilic, homogeneous and sometimes fibrinoid material. There is also mucin deposition throughout the dermis, seen as bluish material between the collagen bundles, a very characteristic finding in lupus erythematosus. The dermis itself may show bluish, purpuric, ischemic-like areas. Focal extravasation of red blood cells in the upper dermis may be seen. In contrast to discoid lupus erythematosus, the lymphocytic infiltrate is rather discrete in SLE, particularly since there is absence of the patchy perivascular and periadnexal infiltrate seen in the chronic cutaneous form (Fig. 6.4).

Bullous Lupus Erythematosus

There is a prominent neutrophilic infiltrate with nuclear dust in the papillary dermis, adjacent to the dermo-epidermal junction and sometimes trecking along the basal layer of the epithelium of hair follicles. Bulla formation ensues. Other areas of the biopsy may show characteristic findings of lupus erythematosus (Fig. 6.5).

Immunoflourescence

The lupus band test is positive in 90% of involved skin and in 80% of noninvolved sun-exposed skin (usually volar aspect of arm) of patients with SLE. Of diagnostic importance is the lupus band test from uninvolved sun-protected skin, usually buttocks, in which a positive lupus band test often implies active renal disease. The lupus band test is demonstrated with direct immunofluorescence and it is characterized by a continuous granular line or band along the dermo-epidermal junction either with IgG alone, or IgG and IgM as well as C3.

Laboratory Abnormalities

The most sensitive test is the ANA, although it is not specific for SLE. Antinuclear antibodies found in SLE include antinative DNA, antidouble stranded DNA, antinucleoprotein antibodies, anti-Sm (Smith) antigen, anti-SS DNA antibodies and anti-nRNP antibodies. Sm antibodies are indicative of significant renal involvement. Anticytoplasmic antibodies RO/SS-A antigen and LA/SS-B antigen are often found in patients with subacute cutaneous LE (SCLE). Other laboratory abnormalities include hemolytic anemia, leukopenia, lymphopenia, and hypocomplementemia.

Dermatomyositis

Clinical

Clinically characterized by polymyositis and skin lesions which include heliotrope rash, poikilodermatous dermatitis, Gottron's papules over the knuckles, telangiectasia of nail folds, and occasionally calcifications. Patients usually have muscle weakness and in 10% of patients there is an associated internal malignancy.

Histology

The histologic findings are those of poikilodermatous dermatitis, including hyperkeratosis, atrophy of the epidermis, hydropic degeneration of the basal layer, telangiectasia of the papillary dermal capillaries, a mild band-like lymphocytic infiltrate, and pigment incontinence. The findings are indistinguishable from those seen in SLE.

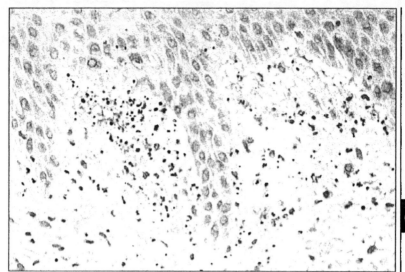

6

Fig 6.5. **Bullous lupus erythematosus**: Interface dermatitis with neutrophils and nuclear dust along the dermoepidermal junction. Other findings of lupus are also seen.

Lichen Sclerosus et Atrophicus (LS&A)

Clinical

The typical location is on perianal or vulvar skin of postmenopausal elderly women. The clinical appearance is that of an atrophic, ivory white plaque with cigarette paper appearing epidermis. With time, it may produce atrophy and sclerosis of the area (**kraurosis vulvae** in women, **balanitis xerothica obliterans** in men). However, it is not uncommon to see LS&A elsewhere on the body, particularly the trunk, and in younger patients. Not uncommonly, plaques of LS&A are indurated and bound-down to the subcutaneous tissue, and in those cases the condition is usually associated with **morphea** (localized sclerosis). There is the opinion by some, that LS&A is a superficial type of **morphea** (localized sclerosis). The histologic findings however, are dissimilar enough to consider them different disorders, even if related.

Histology

The histologic features of LS&A are characterized by hyperkeratosis with follicular plugging, a thin atrophic epidermis with absence of rete ridges, and hydropic degeneration of the basal layer. Characteristically the dermis shows an edematous, pale staining, and homogenous band which involves the papillary dermis and even the most superficial part of the reticular dermis. That area is acellular and devoid of infiltrate, while a band-like lymphocytic infiltrate is seen immediately below (Fig. 6.6). In the case of LS&A/morphea overlap, the deep dermis shows changes characteristic of morphea in addition to the changes described for LS&A.

6

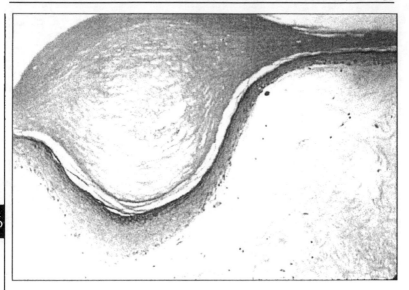

Fig 6.6. **Lichen sclerosus et atrophicus**: Hyperkeratosis with follicular plugging, atrophy of the epidermis and edema/homogenization of the papillary dermis.

Section 13. Necrotic Keratinocytes

Necrotic keratinocytes are seen in interface and lichenoid dermatitis and they are characterized by eosinophilic cytoplasm and absence of nuclei. When the changes are early a picnotic nucleus may still remain in the cell. Dyskeratosis is a term used to describe this phenomenon, particularly when the nucleus is still present, as seen in actinic keratoses and squamous cell carcinoma. The necrosis of epidermal cells can be either isolated or confluent, and is presumably the result of lymphocyte mediated apoptosis.

Erythema Multiforme (EM)

Clinical

The characteristic lesion is the target or bull's eye macule with two or more concentric rings, the central one being often edematous with possible bulla formation, and the outer ring being erythematous. Other lesions including macules, papules and bullae may present on the trunk and extremities, including palms and soles. Mucosal lesions are often seen, particularly in severe cases of erythema multiforme (Stevens-Johnson's syndrome). According to the severity of the clinical involvement EM has been classified as major and minor. It is a reactive process that is usually triggered by viral or bacterial infections, exposure to drugs or associated with neoplastic conditions. Among the infections the two most common are herpes virus type 1 and mycoplasma pneumonia. Among the offending drugs the most common are sulfonamides and nonsteroidal anti-inflammatory drugs.

Table 6.3. Isolated necrotic keratinocytes

Erythema multiforme
Pityriasis lichenoides et Varioliformis acuta
Graft vs host disease
Lichen planus
Benign lichenoid keratosis
Lupus erythematosus
Sun damaged skin/Radiation dermatitis
Some drug reactions (Photodrug)

Table 6.4. Confluent necrosis of epidermal cells

Necrolytic migratory erythema
Acrodermatitis enteropathica
Toxic epidermal necrolysis (TEN)
Ulcerative and febrile PLEVA

6

Histology

The typical findings in erythema multiforme are characterized by a perivascular lymphocytic infiltrate in the upper dermis, with some sprinkling of lymphocytes toward the basal layer of the epidermis. There is often edema of the papillary dermis which on occasion is so pronounced as to form a bulla. The epidermal basal layer shows hydropic degeneration and individual necrotic keratinocytes (dyskeratotic cells) (Fig. 6.7). In older lesions the necrotic keratinocytes may appear in the middle or upper part of the stratum Malpighii, the result of upward mobilization. Occasionally EM may show mainly dermal inflammatory changes and edema, with only minor epidermal involvement (dermal EM). In severe cases, on the other hand, the necrosis of the epidermis can be prominent and become confluent, approaching toxic epidermal necrolysis (TEN). It is unclear if TEN is the end spectrum of EM or if it is a separate entity. We favor the latter, since the clinical presentation and the histologic findings point toward a specific condition; however, there is common ground between both entities (see below).

Necrolytic Migratory Erythema

Clinical

It is the skin manifestation of the glucagonoma syndrome, and it is usually due to a glucagon secreting islet cell tumor of the pancreas. The skin manifestations are characterized by annular erythematous plaques with superficial necrosis and formation of flattened bulla and erosions particularly in the perioral region, trunk, groin, perineal areas, thighs and buttocks. In addition, these patients often have glossitis, glucose intolerance, elevated glucagon levels and decreased plasma aminoacids. Skin lesions usually improve with aminoacid replacement.

Histology

The characteristic histologic finding is the presence of a sharply delineated band of necrosis of the superficial half of the epidermis, including the upper part of the stratum Malpighii and the stratum corneum. The cells immediately below the area of necrosis show a clear, vacuolated cytoplasm and eventually become necrotic, while the keratinocytes in the lower part of the epidermis are normally viable. Subcorneal pustules with crusts, erosions, and yeast overgrowth are found in older lesions.

Acrodermatitis Enteropathica

Clinical

Clinically frequently presents in infants at the time of weaning and is caused by zinc deficiency in individuals genetically predisposed. The disorder is recessively inherited. A similar clinical picture can be seen in zinc deficiency situations brought on by malnutrition, parenteral nutrition deficient in zinc, alcoholic cirrhosis and other conditions that prevent the intake of zinc. A perioral, perianal and acral eczematous eruption, sometimes with vesicles and bullae formation is seen.

Histology

Similar to the findings seen in necrolytic migratory erythema.

Toxic Epidermal Necrolysis (TEN)

Clinical

TEN is usually classified as the end spectrum of severe erythema multiforme. Presently TEN frequently is grouped with Stevens-Johnson's syndrome as erythema multiforme major. There are, however, clinical and histologic features that suggest that TEN may be a specific entity. Clinically it has a rapid onset, is almost exclusively triggered by a drug intake, the most common being sulfonamides, nonsteroidal anti-inflammatory drugs and hydantoins. A rash may precede in which target lesions are sometimes seen. The typical manifestations include flaccid bulla that break easily leaving extensive denuded areas, on occasion involving most of the body. In order to make a diagnosis of TEN, the area involved should be at least 10% of the corporal surface.

Histology

There is separation of the epidermis from the dermis with production of a non-inflammatory, subepidermal bulla. The epidermis shows full thickness necrosis manifested by loss of the tinctorial properties of the nuclei, which appear pale or eosinophilic rather than basophilic, and occasionally become erased. The cytoplasm of the cells is homogeneous and eosinophilic. The histologic findings suggest ischemic necrosis of the epidermis, rather than individual cell necrosis. In contrast to an erythema multiforme, the dermis is completely devoid of inflammatory cells (Fig. 6.8).

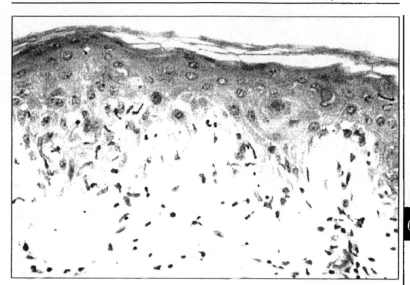

Fig 6.7. **Erythema multiforme**: Interface dermatitis with individual necrotic keratinocytes along the basal layer or throughout the stratum of Malpighii.

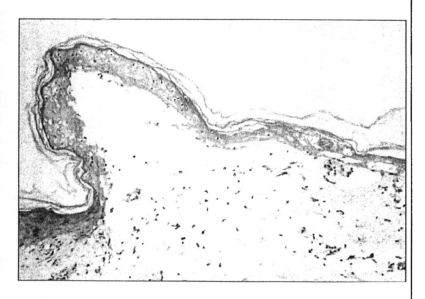

Fig 6.8. **Toxic epidermal necrolysis**: There is complete necrosis of the dermis which appears lifted with subepidermal blister formation.

Ulcerative and Febrile PLEVA

Clinical

This is a severe form of pityriasis lichenoides et varioliformis acuta, characterized clinically by extensive, widespread, varioliform, hemorrhagic and necrotic lesions.

Histology

Histologically there is an interface dermatitis with a marked mixed infiltrate accompanying extensive necrosis of the epidermis, focal epidermal regeneration, and crust formation with occasional secondary yeast overgrowth.

Necrosis of Eccrine Glands

Clinical

Patients with drug induced coma, particularly due to barbiturates, may develop bullae, hemorrhagic or erythematous macules and erosions on dependent or pressure areas.

Histology

Histologically the characteristic finding pointing to the etiology of the condition is necrosis of cells in the secretory acini of the sweat glands, as well as in the eccrine duct. The myoepithelial cells usually survive. The infiltrate in the dermis is sparse with neutrophils and lymphocytes.

CHAPTER 7

Lichenoid Dermatitis, Pigmentary Incontinence and Perivascular Infiltrates

Ramón L. Sánchez and Sharon S. Raimer

Section 14. Lichenoid Dermatitis

Lichenoid infiltrate or lichenoid dermatitis is defined as a band-like infiltrate of lymphocytes running parallel to the epidermis and usually involving the dermo-epidermal junction. Lichenoid dermatitis is a type of interface dermatitis; however, many of the interface dermatitides lack a lichenoid infiltrate.

Lichenoid Dermatitis

Lichen Planus (LP)

Clinical

It is characterized by polygonal, purple, and pruritic papules with white striae (Wickham's striae). It is usually seen on the extensor surface of the extremities, on the genitalia and often on the oral mucosa. Age of involvement is variable, with both children and adults being affected. Erosive lichen planus of the oral mucosa may present without skin involvement and may follow a protracted course. Classic LP usually resolves spontaneously after one or two years. A variant seen in Middle East countries is **lichen planus actinicus**, involving exposed areas of the skin, particularly the face. Oral LP has been associated with hepatitis C.

Histology

There is **hyperkeratosis**, normal to thickened granular layer with wedge hypergranulosis, and irregular acanthosis. The acanthosis has a characteristic "saw tooth" appearance due to remodeling of the basal layer which results in pointed rete ridges. There are abundant necrotic keratinocytes, (dyskeratosis, apoptosis) seen both in the epidermis and as groups of eosinophilic bodies in the papillary dermis. They are termed Civatte bodies or colloid bodies. They stain positive for PAS, measure about 20 microns in diameter and are characteristic, although not exclusively seen, in lichen planus. In the upper dermis there is a band-like infiltrate of lymphocytes **(lichenoid infiltrate)** running parallel to the surface and obscuring the dermo-epidermal junction. These lymphocytes are damaging the basal layer as indicated by the necrotic keratinocytes and the remodeling of the epidermis. The epidermal basal layer shows hydropic degeneration (vacuolar interface dermatitis). Pigment

Table 7.1. Lichenoid dermatitis

Lichen planus
Lichen nitidus
Lichen striatus
Lichenoid drug eruption
Graft vs host disease
Benign lichenoid keratosis
Parapsoriasis variegata
Mycosis fungoides

Table 7.2. Entities with necrotic keratinocytes

Lichen planus
Benign lichenoid keratosis
Fixed drug eruption, photodrug
Erythema multiforme
Graft vs host disease
PLEVA
Lupus erythematosus
Sun damage/radiation dermatitis

incontinence is seen in the dermis (Fig 7.1). Direct immunofluorescence in LP shows immunoglobulin and complement deposition in the colloid bodies.

Variants of Lichen Planus

Bullous Lichen Planus
Clefts are formed between the epidermis and dermis by confluence of the hydropic changes (Max Joseph spaces), with production of bulla.

Hypertrophic Lichen Planus
Often seen on the shins. It resembles prurigo. Histologically there is marked acanthosis, hypergranulosis and a lichenoid infiltrate.

Atrophic Lichen Planus
With thin diminished and flat epidermis.

Lichen Planus Actinicus
Lesions seen on the face and other sun exposed areas, clinically resembling melasma.

Lichen Plano-Pilaris
Keratotic follicular papules associated with scarring alopecia of the scalp, which can become generalized. Histopathologically it shows a lichenoid infiltrate surrounding the follicular epithelium, particularly around the infundibulum and isthmus. The epidermis may show changes of lichen planus. Clinical and histologic differential diagnosis should be made with discoid lupus erythematosus.

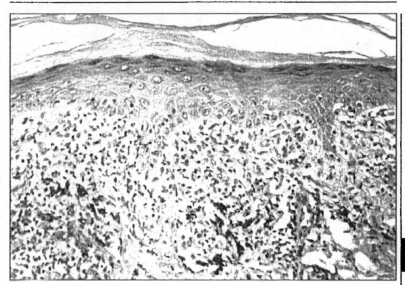

Fig 7.1. **Lichen planus**: Hyperkeratosis, hypergranulosis, saw tooth remodeling of the epidermis and a lichenoid and interface dermatitis with pigment incontinence.

Lichen Nitidus

Clinical
Clinically it is characterized by small, 1-2mm papules usually located on the extensor surface of the extremities, the trunk and genitalia of children or young adults.

Histology
The individual papules are seen as collections of inflammatory cells in the papillary dermis, adjacent to and in continuity with the epidermis. The infiltrate includes lymphocytes, macrophages, occasional plasma cells, melanophages, epithelioid cells and multinucleated giant cells. The overlying epidermis is thin and often shows parakeratosis. Around the edges of the infiltrate the epidermis is hyperplastic forming a claw-like epidermal proliferation which partially embraces the infiltrate (Fig. 7.2).

Lichen Striatus

Clinical
Clinically it is characterized by a linear, papular eruption, usually on the extremities, sometimes involving the full length or rarely following Blaschko's lines. It is usually seen in children and adolescents and as a rule has spontaneous resolution after one or two years.

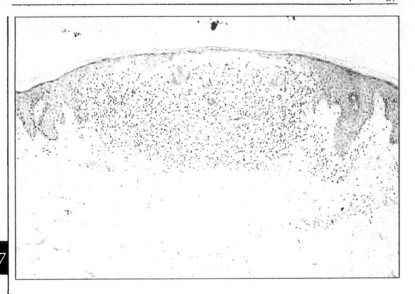

Fig 7.2. **Lichen nitidus**: There is a subepidermal infiltrate of lymphocytes and macrophages with giant cells. The overlying epidermis is thin.

Histology

The epidermis shows spongiotic changes with exocytosis and occasional vesicle formation. Dyskeratotic cells are seen throughout the epidermis. In the dermis there is a perivascular lymphocytic infiltrate that reaches focally the epidermis and becomes lichenoid.

Parapsoriasis Variegata

Clinical

Parapsoriasis variegata shows clinically atrophic, hyperpigmented and irregular patches and plaques.

Histology

Histologically, the epidermis is often atrophic. There is an interface dermatitis with a marked lichenoid infiltrate and prominent pigment incontinence.

Section 15. Pigmentary Incontinence

Pigment incontinence is characterized clinically by hyperpigmented macules, and histologically by the presence of melanophages (macrophages that have phagocyted melanin) which are located around the capillaries of the superficial plexus. Extracellular melanin granules in the dermis are also seen on occasion. Pigment incontinence implies a previous inflammatory reaction involving the basal layer of the epidermis (interface dermatitis), with destruction of melanocytes and dropping

---

Table 7.3. Disorders with pigment incontinence

Interface and lichenoid dermatoses
Erythema dyschromicum perstans
Drug eruption
Postinflammatory hyperpigmentation
Incontinentia pigmenti
Poikilodermatous dermatitis
Erythema ab-igne
Melasma

---

of the melanin into the papillary dermis, where it is phagocytized by macrophages. It is also referred to as postinflammatory hyperpigmentation. It may alternate with postinflammatory hypopigmentation, loss of pigment secondary to inflammatory processes. It is seen after trauma, cryotherapy, infections, eczema, drug reactions, lichenoid dermatoses, and lupus erythematosus. It is more prominent in darkly pigmented individuals.

7

Erythema Dyschromicum Perstans (EDP)

Clinical

Clinically, it is seen mainly in patients with increased natural pigmentation, particularly individuals of Latin American origin. It has also been described as ashy dermatosis (**dermatosis cenicienta**). It is characterized by large blue-gray macules and patches particularly on the face, trunk, arms and thighs.

Histology

Very often the only histologic findings are those of postinflammatory hyperpigmentation, with melanophages in the upper dermis. Therefore it is indistinguishable from chronic or resolved drug eruption and other inflammatory processes resulting in pigment incontinence. Early lesions of EDP show a lichenoid infiltrate with colloid bodies resembling lichen planus (Fig. 7.3). This process has great similarities to **lichen planus actinicus.**

Section 16. Perivascular Infiltrates

Perivascular lymphocytic infiltrates are seen in most inflammatory dermatoses, often as part of other, more specific lymphocytic distributions such as interface dermatitis, lichenoid dermatitis, spongiosis or epidermotropism. When the perivascular infiltrate around the superficial capillary plexus is the only finding, the biopsy is generally nonspecific. If the perivascular infiltrate is both superficial and deep the ream of possibilities is more limited and several suggestions can be offered to the clinician who took the biopsy. Among the superficial, nonspecific perivascular lymphocytic infiltrates we will refer to three conditions: PUPPP, papular acrodermatitis of childhood, and mucocutaneous lymph node syndrome.

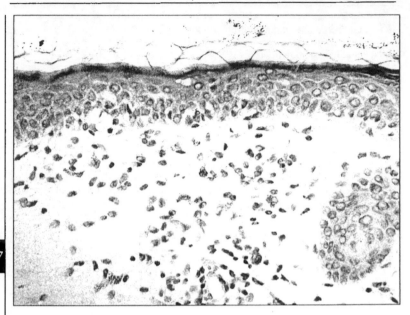

Fig 7.3. **Erythema dyschromicum perstans**: Interface dermatitis with hydropic degeneration and prominent pigment incontinence.

Pruritic Urticarial Papules and Plaques of Pregnancy (PUPPP)

Clinical

Clinically it is seen most commonly in primigravidas, in the third trimester of pregnancy. It is very pruritic and usually starts on the abdomen with extension to the proximal parts of the extremities.

Histology

Shows a superficial and perhaps mid-dermal perivascular lymphocytic infiltrate with variable numbers of eosinophils and neutrophils. Edema of the superficial dermis is often present. The epidermis may show spongiosis, exocytosis and mild acanthosis.

Papular Acrodermatitis of Childhood (Gianotti-Crosti)

Clinical

Clinically erythematous papules, symmetrical, on the face, neck, extremities and buttocks are usually seen in small children. Originally described as a sequel of hepatitis B, now it has been associated with other, more common viral infections including Epstein-Barr virus, coxsackie virus, parainfluenza virus, polio vaccine and cytomegalovirus among others.

Histology

Perivascular lymphohistiocytic infiltrate in upper and mid dermis. Focal spongiosis with exocytosis of lymphocytes may also be seen.

Mucocutaneous Lymph Node Syndrome (Kawasaki's Disease)

Clinical

Clinically it presents with fever and at least four of five clinical criteria including:

1. bilateral conjunctivitis,
2. erythematous oral mucosa with fissured lips and strawberry tongue,
3. erythema and edema of hands and feet,
4. polymorphous skin rash, and
5. cervical nonsuppurative lymphadenopathy.

The rash is usually maculopapular, morbiliform or urticarial. A death of 2% of patients due to cardiovascular complications has been reported. 20% of patients may develop coronary artery aneurysm.

Histology

Skin biopsies show a perivascular lymphohistiocytic infiltrate, papillary dermal edema and dilatation of blood vessels. Exocytosis of lymphocytes in the epidermis may be seen.

Superficial and Deep Perivascular Lymphocytic Infiltrates

Papular Eruption of AIDS

Clinical

A pruritic rash on trunk and extremities with follicular arrangement, seen in AIDS patients.

Histology

There is a marked perivascular superficial and deep lymphocytic infiltrate admixed with numerous eosinophils. Some degree of folliculitis and occasionally granulomas can also be seen.

Erythema Annulare Centrifugum (EAC)

Clinical

It is one of the reactive erythemas, also called **figurate or gyrate erythemas**. Erythema gyratum repens and erythema marginatum are also reactive erythemas. EAC is characterized by annular, occasionally atrophic patches and plaques, often polycyclic, with a slightly elevated outer rim on the trunk and proximal extremities. It is seen sometimes in association with drug ingestion or with minor infection such as tinea pedis but other times can not be related to a known cause. Clinically and histologically EAC can be divided into a superficial and a deep form.

Table 7.4. Superficial and deep perivascular lymphocytic infiltrates

Papular eruption of AIDS
Erythema chronicum migrans
Erythema annulare centrifugum
Erythema gyratum repens
Erythema marginatum
Perniosis (chilblains)

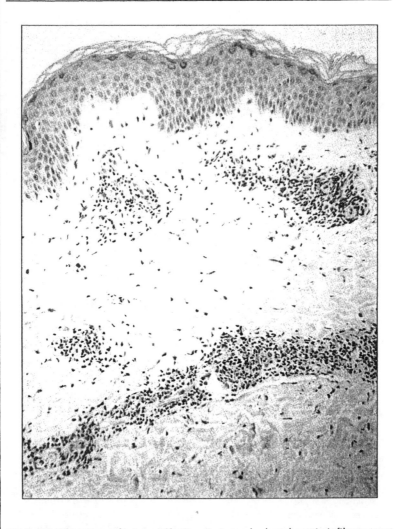

Fig 7.4. **Erythema annulare centrifugum**: Perivascular lymphocytic infiltrate (coat sleeve) in both superficial and deep dermis.

Histology
In the superficial form there is a prominent perivascular lymphocytic infiltrate around the superficial vessels of the dermis. The infiltrate characteristically shows a coat sleeve pattern. In the superficial and deep or indurated type, the infiltrate surrounds not only superficial capillaries but also the deep vessels of the dermis (Fig. 7.4).

Erythema Gyratum Repens

Clinical
This is a very rare condition characterized clinically by annular, concentric and parallel bands of erythema that produce a "wood-grain" pattern on the skin. This condition is associated with internal malignancies in up to 30% of cases.

Histology
This is a perivascular infiltrate, nonspecific, mainly lymphocytic although eosinophils can be also seen.

Erythema Marginatum

Clinical
Clinically seen as red papules and macules particularly on distal aspects of extremities. This condition was seen in association with rheumatic fever, extremely uncommon nowadays.

Histology
There is perivascular lymphocytic infiltrate, nonspecific.

Perniosis (Chilblains)

Clinical
Clinically characterized by tender nodules and plaques on the fingers or toes, after exposure to cold, humid climates.

Histology
There is edema of the papillary dermis and a marked perivascular lymphocytic infiltrate around superficial capillary plexus. In deep lesions, a mononuclear perivascular infiltrate can be seen throughout the dermis and into the subcutaneous fat.

CHAPTER 8

Infiltrates by Lymphocytes, Neutrophils and Eosinophils

Ramón L. Sánchez and Sharon S. Raimer

Section 17. Patchy Lymphocytic Infiltrates

Patchy lymphocytic infiltrate defines a group of conditions in which there is a marked infiltrate throughout the dermis that is arranged in well delineated patches, generally with a perivascular, interstitial, and a periadnexal distribution. It usually involves superficial and deep dermis. Traditionally there were five conditions associated with this type of lymphocytic response (5Ls). A few more have now been added as shown in the table below.

Discoid Lupus Erythematosus (DLE)

Clinical

It is the chronic cutaneous end of the spectrum of lupus erythematosus (see also Chapter VI, section 12). It presents with sharply delimited, hyperkeratotic and erythematous plaques, with prominent follicular plugging. Areas of hyper and hypopigmentation are often seen, particularly in older, atrophic lesions. It is most commonly seen on the face, neck, and upper extremities, particularly the hands. The scalp is frequently involved where DLE produces a scarring alopecia. Although some patients with systemic lupus erythematosus may have discoid lesions, most patients with DLE will never progress to the systemic counterpart. Only a few patients with DLE have elevated ANA titers. Hypertrophic DLE can mimic squamous cell carcinoma both clinically and histologically.

Histology

There is hyperkeratosis and follicular plugging. The epidermis is usually atrophic and shows hydropic degeneration (interface dermatitis) with occasional necrotic keratinocytes. The interface changes are often seen in the follicular epithelium, even when those changes are not present in the epidermis. The lymphocytic infiltrate is perivascular, interstitial and characteristically periadnexal. The lymphocytes are grouped in patches, in the superficial and deep dermis, and the patchy architectural arrangement can be noticed clearly on low magnification examination (Fig. 8.1). Hair follicles, sebaceous glands and eccrine glands are all surrounded and infiltrated by inflammatory cells. The dermis can show also foci of extravasated red blood cells and some degree of edema and/or mucin deposition. Direct immunofluorescence of early

---

Table 8.1. The five Ls, lymphocytic infiltrates

Discoid lupus erythematosus
Jessner's lymphocytic infiltrate
Polymorphous light eruption
Chronic insect bite
Pseudolymphoma (lymphoid hyperplasia)
Lymphoma/leukemia
Syphilis
Lymphomatoid papulosis

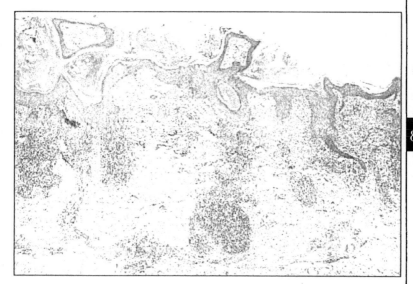

Fig 8.1. Discoid lupus erythematosus: Hyperkeratosis with atrophy of the epidermis. Interface dermatitis and a patchy infiltrate both perivascular and periadnexal.

involved skin generally shows a positive lupus band characterized by granular deposits of IgG, IgM, and sometimes complement along the basement membrane zone. Direct immunofluorescence of uninvolved skin is usually negative in DLE.

Subacute Cutaneous Lupus Erythematosus (SCLE)

Clinical

Clinically it is characterized by annular to polycyclic nonscarring plaques with photo distribution on the face, neck, upper trunk and arms. Mild systemic symptoms such as musculoskeletal complaints may be present. The ANA test is often positive in patients with SCLE, although some patients may have negative ANA with positive anticytoplasmic antibodies to Ro/SS-A or La/SS-B.

Histology

The histologic findings are somewhat intermediate between DLE and SLE, with atrophy of the epidermis and vacuolar interface dermatitis and less infiltrate, although with similar distribution as in discoid lesions.

Chronic Arthropod Bites

Clinical

Occasional arthropod bites, particularly from ticks and chiggers, can persist for several months, presenting as erythematous, pruritic papules with secondary pruritic urticarial papules and plaques of pregnancy-like changes.

Histology

Histologically there is a marked infiltrate throughout the dermis which characteristically is top- heavy (rather than bottom heavy as in lymphomas). The infiltrate is polymorphous with most of the cells being lymphocytes, but plasma cells, histiocytes, eosinophils and neutrophils are seen in variable numbers. The endothelial cells in the infiltrate are usually swollen, adding to the polymorphism of the infiltrate. It is not uncommon to see the formation of round, germinal center-like areas in which larger cells, histiocytes and lymphocytes are prominent, further pointing out the benign nature of the process. Occasionally, the lymphoid cells in arthropod bites are stimulated lymphocytes showing large atypical nuclei and a prominent nucleolus that can be confused with a malignant infiltrate, yet are benign, reactive cells.

Jessner's Lymphocytic Infiltrate

Clinical

Clinically presents with individual papules and plaques on the face, neck or upper torso.

Histology

Histologically a marked lymphocytic infiltrate is seen through the dermis usually with nodular configuration, often surrounding cutaneous adnexae. The cells lack pleomorphism and the infiltrate is polymorphous pointing toward the benign nature of the process.

Polymorphous Light Eruption (PMLE)

Clinical

Two types of lesions in PMLE, papulo-vesicular (superficial) and nodular or plaque (deep) may develop. Both types may coexist. PMLE represents an abnormal reaction to sunlight, and the lesions are seen in sun exposed areas.

Histology

The vesicular lesions show marked edema of the papillary dermis and some degree of spongiosis, with vesicle formation and crust. The lesions are similar to photocontact

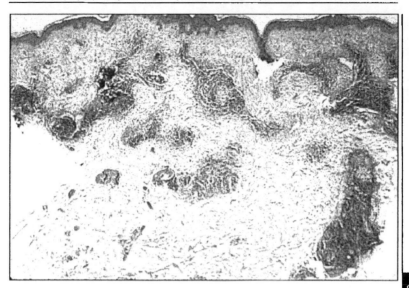

8

Fig 8.2. **Polymorphous light eruption**: Patchy lymphocytic infiltrate, both perivascular and periadnexal without epidermal changes.

dermatitis but there is an absence of sun cells (necrotic keratinocytes in the stratum Malpighii).

The nodular or deep type lesions of PMLE show a patchy lymphocytic infiltrate which is perivascular, periadnexal and indistinguishable from the infiltrates seen in discoid lupus erythematosus. Contrary to DLE, there are no epidermal changes in nodular PMLE (Fig. 8.2).

Pseudolymphoma is better referred to as lymphoid hyperplasia to avoid confusion. Occasionally biopsies of the skin show a marked dermal lymphocytic infiltrate similar to that described for Jessner's or for persistent arthropod bites, that are better classified as lymphoid hyperplasia for lack of clinical correlation.

Lymphoma/Leukemia and sometimes **lymphomatoid papulosis**, exhibit a marked lymphocytic infiltrate of the dermis and therefore they are included in this group of dermal lymphoid infiltrates. Syphilis can also present with a marked lymphoplasmocytic infiltrate of the dermis which justifies the inclusion in this group. For discussion of these entities see Chapters 9 and 22)

Section 18. Diffuse Interstitial Infiltrates

The dermis shows a mild to moderate diffuse infiltrate, which is superficial and deep, without other findings such as vasculitis or folliculitis. Perivascular infiltrates are also seen but in general are discrete. Often the infiltrate is mixed with lymphocytes, neutrophils and eosinophils being the most common cell populations.

1. Insect bites
2. Urticaria
3. Ecthyma, cellulitits
4. Eosinophilic cellulitis (Well's syndrome)

Insect Bites

Insect bites in the acute phase may show a prominent mixed interstitial infiltrate, superficial and deep, in general with large numbers of neutrophils and eosinophils. Depending on the causative arthropod, necrosis of the dermis can also be seen. The differential diagnosis includes urticaria in which there is edema of the dermis and the infiltrate is usually less numerous although often mixed as seen in insect bites (See also Chapter XIII, section 32).

Ecthyma, Cellulitis

The infiltrate is interstitial, superficial, and deep, particularly in the case of cellulitis, and made up mainly of neutrophils. There is usually edema of the dermis. Involvement of the subcutaneous tissue is likely.

Eosinophilic Cellulitis (Well's Syndrome)

Clinically characterized by erythematous, urticarial plaques, particularly on extremities, often measuring 10 or more cm in diameter. Histologically there is a marked interstitial infiltrate with numerous eosinophils, although neutrophils and lymphocytes may also be seen (Fig. 8.3). The eosinophils involve the subcutaneous fat and are found throughout the dermis. Degranulated eosinophils form aggregates of granules surrounded by collagen, forming the characteristic flame figures, which often exhibit basophilic necrosis and peripheral collections of macrophages. Flame figures can be seen in other eosinophilic infiltrates such as insect bites and drug eruption.

Section 19. Neutrophilic Infiltrates

Neutrophils are often seen admixed in small numbers with other inflammatory cells in a variety of skin conditions. In some instances, however, the finding of neutrophils is one of the histologic clues that point toward the diagnosis identity. Neutrophilic infiltrates may appear grouped in so-called microabscesses, or they may be so numerous as to stand out as the main histologic finding. It is helpful to classify the neutrophilic infiltrates of the skin according to location into subcorneal, epidermal, subepidermal and dermal categories.

Subcorneal Pustular Dermatosis (Sneddon and Wilkinson)

Clinical

Clinically presents with recurrent large flaccid pustules particularly on the trunk and intertriginous areas of middle aged patients. Some of these patients have an IgA monoclonal gammopathy or IgA pemphigus.

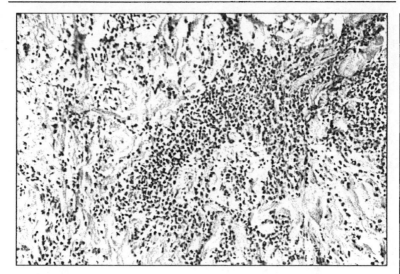

Fig 8.3. Eosinophilic cellulitis: Marked eosinophilic infiltrate admixed with lymphocytes. Flame figures resolved from degranulation of the eosinophils.

Histology

There are large subcorneal pustules filled with neutrophils. Occasionally eosinophils are also seen. The dermis shows a perivascular infiltrate with neutrophils (Fig. 8.4). Subcorneal pustules similar to those just described are sometimes seen following drug exposure to isoniazid diltiazem, cephalosporins, amoxicillin, and anticonvulsants such as hydantoins. When a drug is suspected as the causative agent, this reaction is better classified as a pustular drug reaction.

Impetigo

Clinical

Often seen in children presenting as vesicles or bullae, sometimes complicating pre-existing arthropod bites or eczema. The contagious nature of the process accounts for the multiplicity of lesions in the same patient. *Staphylococcus aureus* is now the most common causative organism. **Group A beta-hemolytic Streptococcus** is sometimes the causative organism. Both organisms may be present in the same lesion.

Histology

Subcorneal pustules with necrotic debris, and fibrin (crust) are seen. There is a neutrophilic infiltrate usually involving the epidermis and the upper part of the dermis (Fig. 8.5). Colonies of bacteria are sometimes seen in the necrotic debris in the crust. Staphylococcal scalded skin syndrome, caused by an epidermolytic toxin produced by some strains of *S. aureus* phage group II, shows histologically a separation of the stratum corneum, but the number of inflammatory cells is sparce.

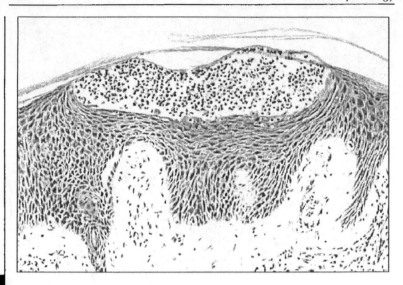

Fig 8.4. **Subcorneal pustular dermatosis**: Neutrophilic pustule in the upper part of the stratum of Malpighii immediately below the stratum corneum.

Table 8.2. Subcorneal neutrophilic Infiltrates

Dermatophyte/Candida infection
Psoriasis/Reiter's syndrome
Seborrheic dermatitis
Subcorneal pustular dermatosis
Pustular drug eruption
Impetigo
Necrolytic migratory erythema
Transient neonatal pustular melanosis
Acropustulosis of infancy
Pemphigus foliaceous

Transient Neonatal Pustular Melanosis

Clinical

It presents at birth with pigmented macules and vesiculopustules particularly on the face, neck and back. The lesions resolve after a few days becoming pigmented macules.

Histology

There are intracorneal or subcorneal neutrophilic microabscesses admixed with fibrin and a few eosinophils. The pigmented macules show increased pigmentation in the basal layer of the epidermis.

Fig 8.5. **Impetigo**: The stratum corneum shows a crust with necrosis, abundant neutrophils and colonies of bacteria.

Acropustulosis of Infancy

Clinical

Clinically characterized by recurrent episodes of extremely pruritic vesiculopustules on the distal extremities of infants. The process resolves at 2 or 3 years of age. This condition may be difficult to differentiate clinically from scabies.

Histology

The histology shows intraepidermal pustules with some eosinophils that eventually become subcorneal. There is a mild perivascular mixed infiltrate in the dermis.

Epidermal Neutrophilic Infiltrates

Neutrophils can be seen diffusely throughout the epidermis in conjunction with subcorneal microabscesses such as in **psoriasis** or **impetigo**, probably as a manifestation of migration of these cells toward the stratum corneum. They are also seen diffusely in the epidermis in **clear cell acanthoma** and in **scabies**. Intraepidermal collections of neutrophils are seen occasionally in **vesicular spongiotic dermatitis**. Intraepidermal neutrophilic microabscesses are also seen in **miliaria rubra**, characterized by collections of neutrophils that, although seen in the epidermis, are located usually inside the acrosyringium. In contrast **milaria crystallina**, commonly seen in children after febrile episodes (heat rash), show intracorneal or subcorneal microabscesses while **miliaria profunda** shows the microabscesses in the lower epidermis or upper part of the dermis. Neutrophilic microabscesses are seen within

Table 8.3. Epidermal neutrophils with pseudoepitheliomatous hyperplasia

Blastomycosis
Chromomycosis
Blastomycosis-like pyoderma
Pemphigus vegetans
Halogen eruptions (Iododerma, bromoderma)

Table 8.4. Subepidermal neutrophilic infiltrates

Bullous pemphigoid
Cicatricial pemphigoid
Dermatitis herpetiformis
IgA bullous dermatosis
Bullous lupus erythematosus
Some drug eruptions

a markedly hyperplastic epidermis in a group of conditions the most characteristic of which is **blastomycosis** (see Chapter 14 section 39). Among those conditions, **blastomycosis-like pyoderma** is characterized by verrucous plaques with pustules and draining sinuses caused by *Staph. aureus*, which histologically show markedly acute inflammation and neutrophilic abscesses in the setting of pseudoepitheliomatous hyperplasia. **Superficial folliculitis (impetigo of Bockhart)** shows collections of neutrophils in the follicular opening, which may resemble intraepidermal neutrophilic collections. In some instances in addition to neutrophils, eosinophils are also seen.

Dermal Neutrophilic Infiltrates

Bullous Lupus Erythematosus

Clinical
It is a subtype of SLE characterized clinically by bullous lesions in addition to other typical findings of this entity.

Histology
Histologically the papillary dermis shows a neutrophilic infiltrate with nuclear dust (**leukocytoclasia**) along the basal membrane zone. As a result, there is separation of the epidermis from the dermis with bulla formation. Other areas of the biopsy may show typical changes of lupus erythematosus.

Sweet's Syndrome (Acute Febrile Neutrophilic Dermatosis)

Clinical
It is seen more commonly in middle aged females and is sometimes recurrent. It is characterized by erythematous papules and plaques with early vesiculation particularly on the extremities, sometimes on the face, less commonly on the trunk. Eruptions

Table 8.5. Dermal neutrophilic infiltrates

Vasculitis (leukocytoclastic)
Arthropod bite
Sweet's syndrome
Pyoderma gangrenosum
Erythema elevatum diutinum
Granuloma faciale
Arteritis
Neutrophilic eccrine hidradenitis
Folliculitis/acne vulgaris
Intestinal bypass syndrome
Deep fungal infection
Papular eruption of AIDS
Reactional leprosy
Excoriation/ulcer
Erysipelas/cellulitis
Ecthyma
Necrotizing fasciitis
Toxic shock syndrome
Granuloma inguinale

8

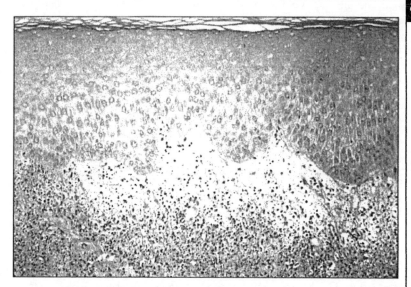

Fig 8.6. **Sweet's syndrome**: There is papillary edema and abundant perivascular neutrophilic infiltrate with nuclear dust. No vasculitis is seen.

are associated with fever, leukocytosis and pain. Ulceration may be the end result. A 10-15% association with leukemia has been repeatedly quoted.

Histology

Early nonulcerated lesions are usually diagnostic and are characterized by a marked perivascular neutrophilic infiltrate particularly in the upper half of the dermis. In association there is extensive nuclear dust (**leukocytoclasia**) but no features of true vasculitis are seen, such as fibrinoid necrosis or extravasation of red blood cells. In addition to the neutrophilic infiltrate the papillary dermis is markedly edematous which accounts for the vesicular appearance of the lesions (Fig. 8.6). Older ulcerated lesions show a mixed inflammatory infiltrate and necrosis which cannot be differentiated from pyoderma gangrenosum.

Pyoderma Gangrenosum

Clinical

It is characterized by the presence of one or more ulcers, which typically present with a purplish undermined border. Early lesions are papules and nodules and they are seen mainly on the lower extremities although the trunk, head and neck are sometimes involved. A **bullous pyoderma gangrenosum** has been described in association with hematological malignancies. In more than half of the cases of pyoderma gangrenosum there is an associated systemic illness such as Crohn's disease, ulcerative colitis, polyarthritis, gammopathy and other conditions.

Histology

There is necrosis of the epidermis and dermis which often extends into the subcutaneous tissue. The inner wall of the ulcer shows necrosis and a mixed infiltrate with abundant neutrophils, while other areas away from the center show a lymphoplasmacytic infiltrate. A lymphocytic vasculitis at the advancing edge of the ulcer may be seen.

Neutrophilic Eccrine Hidradenitis

Clinical

Erythematous nodules and plaques are seen particularly on the trunk. It follows induction chemotherapy for the treatment of cancer, particularly hematologic malignancies.

Histology

Histologically, the secretory component of the eccrine glands show a neutrophilic infiltrate inside the acini, some of which may appear dilated and focally necrotic (Fig. 8.7).

Excoriations

Clinical

Usually superficial, seen in pruritic dermatosis as well as neurodermatitis.

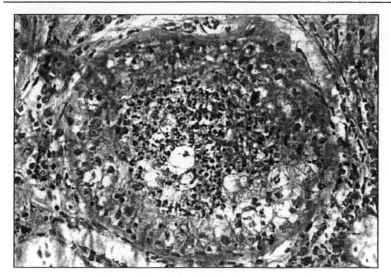

Fig 8.7. **Neutrophilic eccrine hidradenitis**: A eccrine gland is partially necrosed and shows a neutrophilic infiltrate within the gland.

Histology
Histologically there is an absence of epidermis, often with crust formation. The upper dermis shows a neutrophilic infiltrate.

Necrotizing Fasciitis

Clinical
Clinically shows erythema with serosanguineous blisters and necrosis, often on lower extremities. It is usually associated with infection by **group A *Streptococci***. In the scrotum it has been designated as Fournier's gangrene.

Histology
Histologically there is septic vasculitis with necrosis of epidermis, dermis, and subcutaneous tissue, with a mixed inflammatory infiltrate.

Granuloma Inguinale

Clinical
Due to infection by ***Calymmatobacterium granulomatis*** it is clinically seen as ulcers which may cause extensive destruction of the penis or vulva, thighs or lower abdomen.

Histology
Histologically, there is a mixed infiltrate with neutrophilic abscesses, lymphocytes, plasma cells and macrophages around the edge of the ulcer. The diagnostic

feature is the presence of organisms in the cytoplasm of macrophages seen on smears obtained from the ulcer and stained with Wright or Giemsa, the so-called Donovan bodies.

Dermal Lymphocytes

Clinical

Other dermal lymphocytic infiltrates include **toxic shock syndrome**, caused by a toxin produced by some strains of ***Staph. aureus***. It presents with fever, hypotension, redness of mucous membranes, and red patches on the skin.

Histology

Histologically there are foci of spongiosis with neutrophils and a mixed, superficial perivascular and interstitial infiltrate with neutrophils. **Erysipelas** is a form of cellulitis often caused by **group A *Streptococci*** usually seen on the lower extremities or the face. Histologically there is dermal edema, particularly subepidermal, and an interstitial infiltrate with abundant neutrophils. Other forms of cellulitis, while caused by different organisms show similar histologic changes.

Section 20. Eosinophilic Infiltrates

Eosinophils are a component of multiple inflammatory processes, and sometimes they are the key finding to the diagnosis.

Eosinophilic Spongiosis

Clinical

The histological finding is characteristic of early noninflammatory lesions of **bullous pemphigoid** but it can be seen also in **pemphigus vulgaris, pemphigus vegetans,** insect bites, contact dermatitis, **erythema toxicum neonatorum** and **incontinentia pigmentii.**

Histology

It is histologically characterized by the presence of eosinophils in the epidermis, often grouped and associated with spongiosis.

Erythema Toxicum Neonatorum

Clinical

Clinically characterized by 1-3 mm papules or pustules on an erythematous base, although some infants may have only macules, or areas of blotchy erythema, with a peak incidence between 24 and 48 hours after birth in as many as 40% of newborns. The lesions resolve spontaneously, lasting only two or three days.

Histology

The characteristic finding is the accumulation of numerous eosinophils with some neutrophils in the hair follicles. Intraepidermal collections of eosinophils are

---

Table 8.6. Eosinophilic infiltrates

Arthropod reaction, Bite
Scabies
Drug eruption
Hypereosinophilic syndrome
Eosinophilic cellulitis (Well's syndrome)
Eosinophilic fasciitis
Eosinophilic spongiosis
Pemphigus, all types
Bullous pemphigoid
Pemphigoid gestationis
Contact dermatitis
Urticaria
Papular eruption of AIDS
Papular eruption of black men
Erythema toxicum neonatorum
Acropustulosis of infancy
Granuloma faciale
Incontinentia pigmentii
Eosinophilic pustular folliculitis
Eosinophilic ulcer of the tongue

8

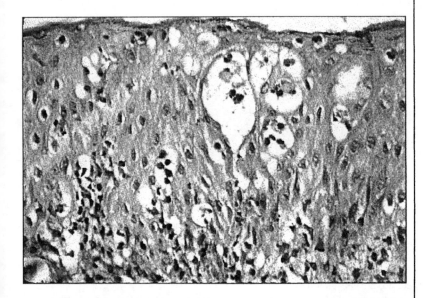

Fig 8.8. **Incontinentia pigmenti**: Vesicular stage. There is spongiosis with exocytosis of the eosinophils (eosinophilic spongiosis) and individual necrotic keratinocytes.

also seen, often located in the follicular opening (**acrotrichium**). The dermis shows also a perivascular eosinophilic infiltrate.

Incontinentia Pigmenti

Clinical

It is a hereditary disorder with an X-linked dominant transmission, and is generally lethal for an affected male fetus. There are three stages. The first is characterized by erythema and bullae arranged in a linear fashion, present at birth or starting shortly thereafter. The extremities are usually involved. The second stage occurs at about two months of life with vesicular lesions being replaced by verrucous plaques, most commonly on extremities and trunk. The third stage is that of swirled hyperpigmentation, generally more prominent on the trunk.

Histology

The first stage is diagnostic and is characterized by **eosinophilic spongiosis** with infiltration of eosinophils throughout the epidermis. **Dyskeratotic, necrotic keratinocytes** are also characteristically seen. The dermis shows a prominent perivascular and interstitial infiltrate of eosinophils (Fig. 8.8). In the second stage there are verrucous lesions with hyperkeratosis, acanthosis, papillomatosis and numerous necrotic keratinocytes. The hyperpigmented areas of the third stage show only deposits of melanin within melanophages in the upper dermis as well as hyperpigmentation in the basal layer.

Eosinophilic Fasciitis (Shulman Syndrome)

Clinical

It is characterized by symmetrical induration of the skin and subcutaneous tissue of the arms and sometimes the legs, with rapid onset. Although it resembles scleroderma or CREST syndrome, there is no involvement of the fingers, no evidence of Raynaud's disease, and no distal involvement. Furthermore, it often resolves after two to four years either spontaneously or following steroid therapy. Of historical interest, cases of eosinophilic fasciitis were recorded after the toxic oil syndrome occurred in Spain, and similar changes with the eosinophilic- myalgia syndrome were noted in the U.S.A. and Japan in 1989 following the ingestion of altered L-tryptophan.

Histology

There is a marked thickening of the deep fascia and septa of subcutaneous tissue with fibrosis and hyalinization of the collagen. The subcutaneous tissue shows edema and infiltration by lymphocytes, plasma cells and eosinophils. Eosinophils are seen also, although not in large numbers, in the fascia. In some cases the sclerotic changes extend into the lower dermis.

Eosinophilic ulcer of the tongue is a chronic ulcer probably caused by trauma on the lateral aspect of the tongue. The histologic findings include a prominent eosinophilic infiltrate in addition to granulation tissue, and acute and chronic inflammation.

CHAPTER 9

Infiltrates by Plasma Cells, Mast Cells and Histiocytes

Ramón L. Sánchez and Sharon S. Raimer

Section 21. Plasma Cell Infiltrates

Plasma cell infiltrates are not a common component of inflammatory dermatosis and their presence should always be noted. Traditionally plasma cells, although not exclusively seen in syphilis, have always been a flag indicating the need for serologic tests for syphilis in the patient and/or silver stains (such as Warthin Starry, or Steiner) for spirochetes. This rule does not apply to inflammatory lesions in or near mucosal orifices, which are often rich in plasma cells. The inflammatory reactions to skin tumors on the face also show an abundance of plasma cells.

Syphilis

Clinical

Infectious disease, usually sexually transmitted, caused by the spirochete *Treponema pallidum*. Classically it is divided into four stages. Primary syphilis or primary chancre, is characterized by an indurated ulcer on the penis, vulvar/vagina, perianal, and occasionally on extragenital areas. In this stage the diagnosis of the condition can be established easily by examination of the exudate of the ulcer with dark field microscopy, which allows visualization of the spirochetes. Secondary syphilis, four to eight weeks after the primary chancre, usually maculopapular, scaly patches or plaques characteristically involving palms and soles, extremities, trunk, and elsewhere. The clinical appearance at this stage is usually very variable and can mimic many other conditions. **Condyloma lata**, verrucous lesions, fleshy and moist, usually perianal but occasionally perilabial or perinasal may be seen. **Alopecia syphilitica** can also be seen in this stage. Necrotic, pustular lesions may be seen, particularly in immunosuppressed patients. Latent syphilis occurs when the secondary stage subsides spontaneously without treatment. This stage may last for years and give abnormal serologic tests. Tertiary syphilis appears many years after the primary infection involving the central nervous system, the cardiovascular system and the skeleton. The skin may show nodules and ulcers.

Histology

The primary chancre shows an ulcer surrounded by reactive, acanthotic epidermis. The base of the ulcer shows a mixed infiltrate with lymphocytes and numerous plasma cells. The plasma cells surround blood vessels which characteristically show

Table 9.1. Plasma cell infiltrates

Syphilis
Yaws/Pinta
Zoom's ballanitis/vulvitis
Secondary process on face or near mucosas
Erythema chronicum migrans
Acrodermatitis chronica atrophicans
Plasmocytoma, multiple myeloma
Waldenstrom's macroglobulinemia
Morphea, inflammatory phase
Lymphoid hyperplasia of the skin

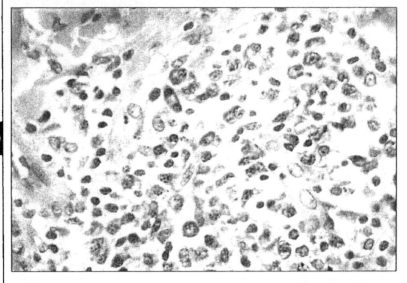

Fig 9.1 . **Secondary syphilis**: The dermis contains a predominantly perivascular infiltrate with abundant plasma cells and edema of the endothelial cells.

endothelial swelling. The treponemes (*Treponema pallidum*) usually can be visualized easily in this stage with double silver impregnation techniques such as Warthin-Starry and Steiner stains. The spirochetes appear as thin, spiral organisms, measuring 4-15 microns in length and to 0.5 microns in diameter. Secondary syphilis shows a dermal, predominantly perivascular, infiltrate with numerous plasma cells. The endothelial cells are prominent (Fig. 9.1). The lesions of **condyloma lata** show marked acanthosis with spongiosis of the epidermis and exocytosis of lymphocytes and plasma cells. The dermis shows similar changes as described above. Usually Warthin-Starry stain yields positive results in condyloma lata, with the organisms usually seen in the areas of spongiosis; on the other hand, it may be difficult to visualize the organisms in secondary syphilis lesions except in cases of immunosuppression. The histologic variations of secondary syphilis includes a prominent dermal infiltrate that can mimic lymphoma, a pustular syphilis with epidermal and dermal neutrophilic abscesses, and a granulomatous syphilis with compact, non-necrotizing granulomas throughout the dermis. Syphilis is often seen in AIDS patients. In alopecia syphilitica a marked lymphoplasmocytic infiltrate surrounds hair follicles some of which appear destroyed.

Tertiary Syphilis

The characteristic findings in this stage is the presence of non-necrotizing granulomas with lymphoplasmocytic infiltrate. No organisms are seen with special stains in this stage. The serologic diagnosis of syphilis is made using the RPR and the FTA (fluorescent treponemal antibodies) tests. The RPR usually becomes positive shortly after the infection, but the titers decrease or even become negative after successful treatment of the disease. A prozone dilution test may be necessary in highly positive patients. The FTA becomes positive after an antibody response has been mounted, usually several weeks after primary infection, and persists elevated indefinitely.

Yaws/Pinta

Clinical

These two conditions are infectious, contagious and nonvenereal. Yaws is produced by *Treponema pertenue* while pinta is due to *Treponema carateum*. Both organisms are indistinguishable from *Treponema pallidum*. The two entities are seen in tropical or subtropical areas, with pinta being localized mainly in the Caribbean and Central America. Yaws starts in childhood and goes through three stages, the first two being papillomatous or verrucous and the last one being characterized by ulcers and gumma. Pinta on the other hand, is characterized by maculo-squamous lesions in the first two stages, which display a large variety of colors, and a third stage with hypopigmented areas.

Histology

The epidermis is hyperplastic, particularly in yaws, and shows marked spongiosis and exocytosis of neutrophils with neutrophilic microabscesses. Similar findings are seen also in pinta although the acanthosis is less prominent. The dermis shows a mixed infiltrate with lymphocytes, plasma cells, and macrophages. The causative organisms are seen with Warthin-Starry stain, particularly in the epidermis.

Erythema Chronicum Migrans

Clinical

It is characterized by erythematous annular plaques that spread peripherally for a period of weeks. It follows a tick bite with introduction of the spirochete *Borrelia burgdOrferi*. The condition, **Lyme disease**, involves also joints, nervous system and heart. The tick vector is *Ixodes dammini* particularly in the northeast United States (Connecticut). *I. ricinus* is the vector in Europe for this condition.

Histology

There is a superficial and deep perivascular and interstitial infiltrate composed of lymphocytes, plasma cells, eosinophils and some macrophages. In early lesions the *Borrelia burgdOrferi* organisms can be seen with silver stains (Warthin-Starry, Steiner), particularly in the upper dermis.

Acrodermatitis Chronica Atrophicans

Clinical

This condition seen mainly in Europe, particularly northern and central, follows infection by a subtype of *Borrelia burgdOrferi*, the **Vs461** subtype also known as *B. burgdOrferi afzelius*. The spirochete is introduced by the tick bite of *Ixodes ricinus*. Clinically the lesions show an early inflammatory stage, particularly on extensor surface of extremities, followed by an atrophic stage with loss of appendages and hypopigmentation. Other areas show fibrous, sclerotic plaques that may resemble lichen sclerosus et atrophicus.

Histology

There is a superficial and deep dermal infiltrate composed of lymphocytes, plasma cells and some histiocytes.

Zoom's Balanitis/Vulvitis (Balanitis Circumscripta Plasma Cellularis)

Clinical

Sharply defined red patch, particularly seen in uncircumcised individuals in the area of the glans and foreskin. Similar lesions can be seen rarely on vulva, and on lips.

Histology

There is a dense, band-like plasmocytic infiltrate in the upper dermis and extending to the basal layer of the epidermis. The latter shows atrophy and edema with separation of keratinocytes which are still attached by the desmosomes, acquiring a polygonal shape (lozenges). Other cells including lymphocytes, neutrophils and eosinophils are sometimes seen admixed in the infiltrate (Fig. 9.2).

Section 22—Mast Cell Infiltrates

The entities related to increased numbers of mast cells can be classified as **cutaneous mastocytosis, systemic mastocytosis** when the infiltrates are in other organs with or without skin involvement, and malignant mast cell disease. The cutaneous mastocytosis are often seen in childhood and usually have good prognosis, generally resolving in puberty. On the other hand, cutaneous mastocytosis in adults are associated with systemic involvement in up to 40% of cases.

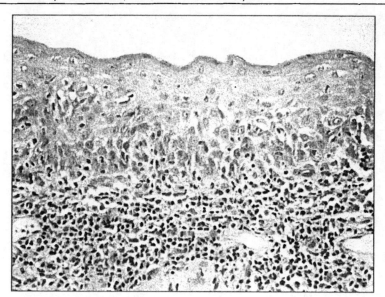

Fig 9.2. **Zoom's balanitis**: Lichenoid infiltrate with lymphocytes and plasma cells. The keratinocytes appear separated by edema (lozenger appearance).

Urticaria Pigmentosa

Clinical

This is the most common of the cutaneous mastocytoses and it is clinically characterized by multiple hyperpigmented papules particularly on the trunk and extremities. The papules tend to urticate, wheel and flair when rubbed (Darier's sign).

Solitary Mastocytoma

Clinical

It is seen usually in childhood as a solitary lesion on the trunk or on the wrist. The individual lesions can be small hyperpigmented papules or nodules.

Diffuse Cutaneous Mastocytoma

Clinical

It usually starts in childhood and is characterized by a diffuse involvement of the skin which appears yellowish brown, or erythematous in color. Vesiculation with bulla formation is common in this subtype.

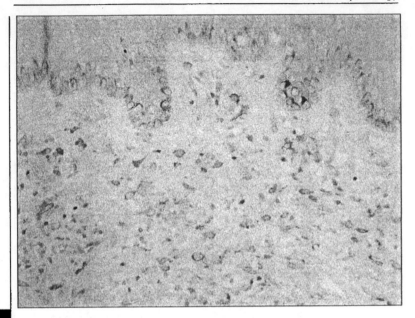

9

Fig 9.3. **Urticaria pigmentosa**: Among the cells in the upper dermis are abundant mast cells that stain metachromatically with Giemsa stain.

Telangiectasia Macularis Eruptiva Perstans (TMEP)

Clinical

This form of cutaneous mastocytosis is seen mainly in adults and is characterized by the presence of hyperpigmented macules with erythema and telangiectasia particularly on the trunk.

Histology

It is similar for all types of cutaneous mastocytosis although it is variable depending on the numbers of cells present in the infiltrate. In **urticaria pigmentosa** the cells are usually located in the upper part of the dermis in a diffuse or band-like distribution (Fig. 9.3). In **TMEP** the number of mast cells can be only slightly increased over normal and they are mainly distributed in a perivascular arrangement. In **cutaneous mastocytoma** there are large numbers of mast cells that usually fill up the dermis. While the mast cells in the other forms of cutaneous mastocytosis are usually elongated or spindled, in cutaneous mastocytoma they become cuboidal or rounded with their nucleus being in the center of the cell (Fig. 9.4). In diffuse cutaneous mastocytosis the mast cells are seen throughout the dermis however the numbers are not so large as in solitary mastocytoma. The mast cells in cutaneous mastocytosis are normal in appearance, they contain cytoplasmic granules that can be demonstrated metachromatically by using giemsa, or toluidine blue stains. They

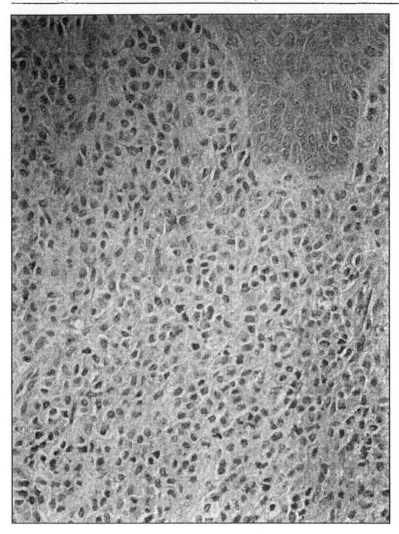

Fig 9.4. **Cutaneous mastocytoma**: A prominent collection of mast cells fills the papillary and reticular dermis. The mast cells are rounded with a central nucleus.

can also be seen with Leder stain (chloroacetate esterase). Presently there are also some immunoperoxidase stains specific for mast cells such as CD117.

Systemic mastocytosis is seen in approximately 40% of adults with cutaneous mastocytosis, including TMEP, while in children only the diffuse cutaneous mastocytosis is usually associated with systemic involvement. Systemic mastocytosis is characterized by involvement of organs other than the skin. The organs that are involved are mainly bone marrow, liver, spleen, gastrointestinal tract and lymph nodes. Systemic mastocytosis may on occasion evolve into malignant mast cell disease.

Malignant mast cell disease can be seen as a proliferation of malignant mast cells in organs or as mast cell leukemia with circulating neoplastic mast cells.

Section 23. Histiocytic Infiltrates

The histiocytoses are a group of diseases with localized or generalized proliferations of cells of the monocyte-macrophage and/or dendritic cell systems. Apart from the rare true malignant histiocytosis, the histiocytoses do not fulfill the criteria for malignancy but rather appear to result from abnormal or altered regulation of histiocyte activity.

Langerhans Cell Histiocytosis/Dendritic Cell Histiocytoses

Clinical

Langerhans cell histiocytosis (LCH) is thought to be a reactive condition in which cells with the phenotype of Langerhans cells invade and damage one or more organs or tissues. Eosinophilic granuloma (localized lesions of bone), Hand-Schuller-Christian disease (the triad of exophthalmos, diabetes insipidus and multiple bone lesions), and Letterer-Siwe disease (skin and visceral lesions) were grouped under the term "histiocytosis X" by Liechtenstein who recognized their pathological similarities. Clinical findings of LCH range from chronic asymptomatic single organ involvement to a multi-system disease with fever, malaise, failure to thrive, organ failure and death. The majority of systemic cases occur between the ages of 1 and 15 years with the peak incidence under age 2 years. Skin lesions of LCH may present as small erythematous papules which may be generalized in distribution, scaling, crusted lesions resembling seborrheic dermatitis or Darier's disease. They frequently occur and tend to be most prominent on the scalp, postauricular, perineal and axillary areas. Nodular lesions which may crust or erode may be present, and infiltrated plaques with a tendency to ulcerate may develop particularly in the groin and flexural areas.

Multiorgan involvement frequently manifests as lymphadenopathy which is occasionally massive, and hepatosplenomegaly. Diabetes insipidus occurs in 25-50% of children with multisystem disease and growth hormone deficiency may result from hypothalamic involvement.

Histology

The site and age of a lesion influence the histologic findings. Lesions examined early show an upper dermal and junctional accumulation of large histiocytic cells with homogeneous pink cytoplasm and lobulated, bean or kidney shaped nuclei. Histiocytic cells may be observed singly or forming Pautrier-like microabscesses within the epidermis. Mitoses are rare and phagocytosis is exceptional. The numbers of eosinophils, macrophages, lymphocytes, and plasma cells are variable. Multinucleated giant cells are frequently found. Over time a more xanthomatous pattern arises which is eventually followed by fibrosis (Fig. 9.5).

Table 9.2. The histiocytoses

Class I	**Dendritic Cell Histiocytoses**
	Langerhans' cell histiocytosis
	Indeterminate cell histiocytoma
Class II	**Histiocytes of mononuclear phagocytes other than Langerhans' cells**
	Juvenile xanthogranuloma
	Benign cephalic histiocytosis
	Eruptive histiocytoma
	Xanthoma disseminatum
	Hemophagocytic lymphohistiocytosis
	Sinus histiocytosis with massive lymphadenopathy
	Multicentric reticulohistiocytosis
Class III	**Malignant histiocytic disorders**
	Monocytic leukemia
	Malignant histiocytosis
	True histiocytic lymphoma
	Malignant Langerhans cell histiocytosis

9

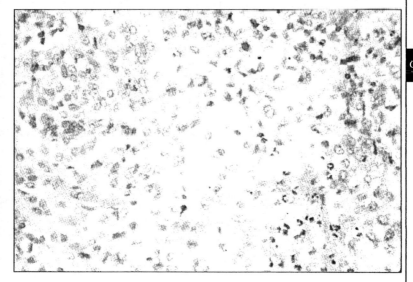

Fig 9.5. **Langerhans cell histiocytosis**: Numerous Langerhans cells in the dermis with bean shaped nucleus. Neutrophils, lymphocytes and eosinophils are also seen.

A congenital nodular form of LCH described by Hashimoto and Pritzker (congenital self-healing reticulohistiocytosis) may differ histopathologically from the above description. The infiltrate may be found deep within the dermis with sparing of the papillary dermis and epidermis and consists of large histiocytic cells with glassy eosinophilic cytoplasm.

Ultrastructurally rod to racket shaped Birbeck granules are found in LCH cells. The Birbeck granule is resistant to destruction by formalin fixation and routine paraffin block embedding and thus can be demonstrated in such material.

Immunohistochemical studies are important in confirming a diagnosis of LCH. Two markers, S-100 protein and peanut agglutin (PNA), are demonstrable in formalin-fixed, paraffin-embedded tissue. LCH cells stain positively with anti-CD1 (OKT6) and anti-CD4 (OKT4) antibodies and histochemical staining shows positive activity for ATPase and for a-mannosidase.

Juvenile Xanthogranuloma (JXG)

JXG is a benign histiocytic tumor which occurs most commonly in infancy and early childhood and tends to be self healing. The pathogenesis of this tumor is unclear.

Clinical

The lesions of JXG generally erupt as one or several small erythematous papules which develop into 2-20 mm (most commonly 5-10 mm) yellowish red nodules.

Lesions of JXG occur most commonly on the upper part of the body. Ocular involvement is the most common extracutaneous manifestation, and lesions affecting the iris may lead to anterior chamber hemorrhage and glaucoma.

Histology

Early lesions show a diffuse infiltration of spindle shaped fibrohistiocytic cells. As the lesion matures, a mixed cellular infiltrate with histiocytes, lymphocytes, eosinophils, and occasional neutrophils and plasma cells is seen. Touton giant cells, characterized by the wreath like arrangement of multiple nuclei are typically present. In older lesions foamy, lipid laden histiocytes appear, and resolution is marked by gradual replacement by fibrous tissue (Fig. 9.6). The lesions commonly extend from just below the epidermis into the subcutaneous fat and are characteristically well delimited from uninvolved dermis.

Immunohistochemical studies have demonstrated that the histiocytic cells in the infiltrate of JXG express CD68 macrophage markers and are negative for peanut agglutin and S-100 protein except for an occasional S-100 positive dendritic cell at the periphery of the lesion. Ultrastructurally lesional histiocytes have complex interdigitations of the cytoplasmic membrane but no Birbeck granules.

Benign Cephalic Histiocytosis

Clinical

The lesions of benign cephalic histiocytosis most commonly appear in the second half of the first year of life as yellow-red 2-5 mm papules on the cheeks that spread to involve the forehead, ears, neck and occasionally the upper body and buttocks. The lesions generally do not increase in size and heal after several years leaving flat or atrophic pigmented scars.

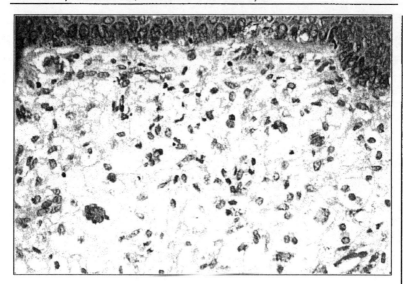

Fig 9.6. **Juvenile xanthogranuloma**: Mixed infiltrate with histiocytes, Touton giant cells, lymphocytes, neutrophils and eosinophils.

Histology

The histology of early lesions shows a well circumscribed infiltrate in the superficial and mid dermis composed primarily of histiocytes with some pleomorphism and scattered lymphocytes and eosinophils. Older lesions may contain a few multinucleated giant cells, but the histiocytes contain no lipids (Fig. 9.7). The cells are S-100 negative. Ultrastructurally, approximately 20% of the histiocytes contain clusters of comma-shaped bodies, but fatty droplets and Birbeck granules are absent. As the condition is self-limiting, no therapy is indicated.

Generalized Eruptive Histiocytoma

Clinical

Generalized eruptive histiocytoma is rare and has been reported more frequently in adults, but also occurs in children.[41] The skin lesions consist of fine papules that are yellow to bluish red in color and range in size from 3-10 mm. The lesions tend to be symmetrically distributed on the face, trunk and proximal limbs, and old lesions tend to fade as new ones appear.

Histology

Histologic examination of the papular lesions reveals a monomorphous histiocytic infiltrate in the upper and mid dermis. A few lymphocytes can be seen but no lipid laden cells or giant cells are observed, which differentiates the lesions from juvenile xanthogranulomas. Stains for S-100 are negative. Ultrastructural studies

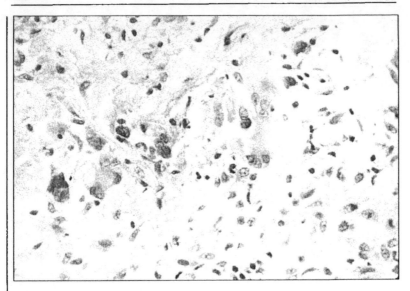

Fig 9.7. **Benign cephalic histiocytosis**: The dermis contains histiocytes and multi-nucleated giant cells with lymphocytes. No foamy histiocytes seen.

have shown that the histiocytic cells lack Birbeck granules but have cytoplasmic laminated bodies.

Xanthoma Disseminatum

Clinical

Xanthoma disseminatum is a rare sporadic histiocytic disease that occurs in both children, predominately older boys, and adults. Skin lesions are typically red-brown to yellow in color and tend to be most prominent on flexural surfaces where they develop in clusters and may become confluent. The eyelids and conjunctiva may be involved, and the lips, pharynx, and larynx characteristically become infiltrated, occasionally resulting in respiratory difficulty. Diabetes insipidus, which occurs in up to 50% of patients, is the only characteristic sign of systemic disease.

Histology

Histopathologically, lesions are similar to those of juvenile xanthogranuloma showing the presence of foamy histiocytes, inflammatory cells, and Touton giant cells. Lesional histiocytes do not bind S-100 or anti-CD1 (OKT6) monoclonal antibodies and are negative for Birbeck granules by electron microscopy.

Hemophagocytic Lymphohistiocytosis (Hemophagocytic Syndrome)

Clinical

The disease occurs most commonly in immunocompromised patients and is usually associated with either an infectious organism, most often a virus, or a neoplastic lymphoid proliferation. The neoplastic lymphoid proliferations are usually of T-cell origin and occur as a familial syndrome in young children 60% of whom are under 6 months of age.

Clinical manifestations may include fever and severe constitutional symptoms, hepatosplenomegaly, lymphadenopathy, coagulopathies, and pancytopenia. Skin manifestations occur in approximately 20% of patients, most commonly as panniculitis or purpura.

Histology

Skin biopsies show erythrophagocytosis and frequently cytophagocytosis of other bone marrow-derived elements by cytologically benign histiocytes. Infiltration of vessel walls and fat necrosis is generally present. Focal dermal necrosis associated with benign and malignant lymphoid infiltrates is common. Erythrophagocytotic histiocytes stain positive with KP-1 and negative for S-100 protein.

Diagnosis and Differential Diagnosis

The diagnosis of hemophagocytic lymphohistiocytosis is based on the histopathological finding of erythrophagocytosis by benign histiocytes. A few true malignancies such as histiocytic lymphomas and malignant histiocytosis, some T cell lymphomas, acute myoblastic leukemia, and myelomas have been reported to exhibit erythrophagocytosis but is usually not a prominent finding and the cells in these conditions are cytologically malignant.

Xanthomas

Clinical

Xanthomas are accumulations of foamy macrophages in the skin and/or subcutaneous tissue due to hyperlipidemia. Depending on the lipoproteins involved, there are different clinical manifestations which allow a classification of xanthomas in five main types, as follows.

Xanthelasma are small yellow papules usually seen on eyelids of elderly patients. They are often associated with senile arc. When xanthelasma appears in younger patients, a hypercholesterolemia should be suspected. **Eruptive xanthomas** are often seen in diabetics associated with high levels of triglycerides. They are small yellowish papules on an erythematous base seen mainly over extensor surfaces such as buttocks, elbows and posterior thighs. **Tuberous xanthomas** are seen mainly in familial hypercholesterolemia and other inherited hyperlipidemias and are characterized by large nodules or plaques often overlying joints. **Tendon xanthomas** are also seen in inherited hyperlipidemias and involve predominantly Achilles tendons

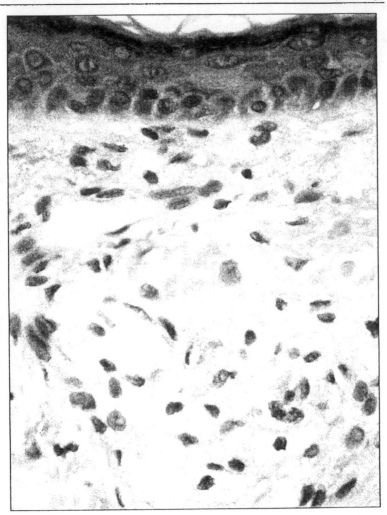

Fig 9.8. **Xanthelasma**: Collection of foamy histiocytes in the upper dermis. Touton giant cells can also be seen.

as well as tendons on the hands. **Plain xanthomas** develop on skin folds and in palmar creases. Among other conditions they may be seen in biliary cirrhosis.

Histology

The histologic hallmark in all xanthomas is the presence of foamy macrophages with small regular nuclei and abundant cytoplasm located throughout the dermis, subcutaneous tissue and/or tendon, depending on the type of xanthoma. Xanthelasma and plain xanthomas are more superficial and the histiocytic infiltrate is located in the superficial half of the dermis (Fig. 9.8). Eruptive xanthomas show infiltrates throughout the dermis and characteristically show additional foamy macrophages, pools of amorphous, bluish material resembling mucus substances. Tuberous xanthomas show large collections of foamy histiocytes throughout the dermis and subcutaneous tissue. Tendon xanthomas show foamy macrophages involving tendon.

Differential Diagnosis

Other tumors show foamy macrophages including cholesterolotic dermatofibroma, juvenile xanthogranuloma and several of the normolipemic histiocytic infiltrates (see above). Sebaceous tumors show foamy cells admixed with epithelial cells in different proportions. Lepromatous leprosy shows foamy macrophages similar to those in xanthomas. Balloon cell nevus and balloon cell melanoma show cells with abundant clear cytoplasm although close inspection shows absence of microvesicles. Likewise, granular cell tumor shows cytoplasmic granules rather than microvesicles. Some lipomas, liposarcomas, and hybernomas also show foamy macrophages.

9

CHAPTER 10

Granulomas

John R. Simmons

Section 24. Giant Cells

Giant cells are large multinucleated cells, usually histiocytes. Keratinocytes, nevus cells, or neoplastic cells may also become multinucleated. Most histiocytic giant cells are nonspecific, although some have a unique histology and accompany characteristic diseases, such as the epithelial giant cells with characteristic nuclear inclusions that are the histologic hallmark of Herpes infection.

Foreign Body Giant Cells
Multiple nuclei haphazardly arranged throughout the cytoplasm. Seen in all types of foreign body reactions as well as some tumors.

Langhans Giant Cells
Nuclei arranged in "horseshoe" pattern at periphery. More common with infectious causes, including tuberculosis.

Touton-type Giant Cells
"Wreath" of nuclei surrounded by foamy cytoplasm. Seen in xanthomas and foreign body reactions. Prototypical tumor is juvenile xanthogranuloma.

Other Giant Cells
Herpes giant cells: seen in epidermis in both biopsy material and in Tzanck smears. Have characteristic nuclear inclusions with "3 Ms": Multinucleation with Molding of nuclei and Margination of chromatin.

Nevus giant cells: typically seen in benign nevi, but may also be seen in atypical melanocytic lesions.

Floret-type giant cells: characteristic appearance, seen in pleomorphic lipoma.

Section 25. Granulomas

Granulomas are collections of histiocytes in the dermis or subcutaneous tissue. Giant cells are often a conspicuous component of granulomas, and while nonspecific, can be useful clues, especially when the granuloma is not readily evident. Granulomas can be classified into three types—epithelioid,(non-necrotizing or naked and necrotizing), foreign body, and necrobiotic—depending upon the accompanying inflammation and dermal changes.

Dermatopathology, edited by Ramón L. Sánchez and Sharon S. Raimer. ©2001 Landes Bioscience.

Table 10.1. Non-necrotizing granulomas

Asteroid and Schaumann bodies seen in:
Sarcoidosis
Tuberculosis
Leprosy
Berylliosis
Foreign body reaction

Sarcoidosis

Clinical

Systemic disease of unknown etiology. May involve any organ g cutaneous lesions in 10-35% (may be only manifestation). Brown-red or purple papules and plaques. **Lupus pernio** refers to lesions on nose, cheeks, and ears.

Histology

Circumscribed epithelioid cell granulomas of varying sizes with little or no inflammation (naked granulomas). Present in dermis or subcutis. May contain few giant cells (usually Langhans type). Giant cells may contain asteroid bodies stellate eosinophilic inclusions, or Schaumann bodies round, laminated, calcific inclusions. Neither is specific for sarcoidosis (Fig. 10.1).

Other Findings

Chest X-ray shows hilar adenopathy (70%). Symptoms depend on organs involved. Kviem test for sarcoidosis, intradermal injection of antigen from sarcoidal tissue, biopsy six weeks later shows sarcoid-like granuloma if positive. Not widely used. Most systemic sarcoidosis diagnosed with bronchoscopy and biopsy.

10

Differential Diagnosis

Sarcoidosis is a diagnosis of exclusion, and stains for mycobacteria (Fite) and fungi (PAS) should be performed, as well as polarization for foreign material. Negative stains do not necessarily exclude other diagnoses, especially with early tuberculous granulomas. All entities in this chapter should be considered before sarcoidosis is diagnosed.

Granulomatous Rosacea

Clinical

Persistent erythema and telangiectasia with papules and pustules on the cheeks, chin, nose, and forehead. Onset usually after 30. Probable multifactorial etiology.

Histology

Small epithelioid cell granulomas with or without focal necrosis and mixed inflammatory infiltrate around vessels or pilosebaceous units. Epidermis may be normal or atrophic (Fig. 10.2).

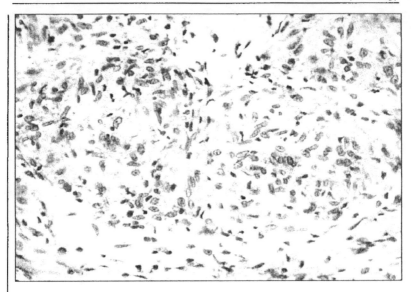

Fig 10.1. **Sarcoidosis**: Non-necrotizing granulomas made up predominantly by epithelioid cells with few lymphocytes (naked granuloma).

10

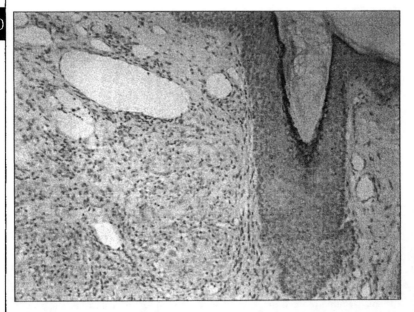

Fig 10.2. **Granulomatous rosacea**: Non-necrotizing granulomas are seen in the superficial dermis with perifollicular arrangement.

Differential Diagnosis

Sarcoidosis, Lupus vulgaris. Perifollicular granulomas and marked vascular dilatation favor rosacea.

Leprosy (Hansen's Disease)

Clinical

Chronic infection by *Mycobacterium leprae*. The organism has low infectivity and a long incubation period (average 5 years). It affects skin, nasal mucosa, and peripheral nerves. Leprosy infection can be divided into two subtypes, depending on host immunity. **Tuberculoid leprosy** results from a strong T-cell response and is a relatively stable form. Clinically, lesions are well-defined red-brown anesthetic plaques on trunk and limbs. Enlarged nerves may be noted. **Lepromatous leprosy** results from deficient T-cell immunity and a poor host response. Cutaneous lesions are symmetrical, hypoesthetic, and may be macular, infiltrative-nodular, or diffuse. Facial involvement may give leonine facies. **Reactional leprosy** can be a reversal reaction or a downgrading reaction. Erythema nodosum leprosum is a type 2 reversal reaction that shows histologically a leukocytoclastic vasculitis.

Histology

Tuberculoid leprosy has well-formed granulomas along neurovascular bundles with dense peripheral lymphocytic infiltrate. Acid-fast bacilli are rarely found with Fite stain, even in nerves (Fig. 10.3). Lepromatous leprosy typically has a grenz zone, separating the epidermis from an extensive dermal infiltrate of macrophages stuffed with bacilli. Because of the poor host response, the macrophages are not activated to form granulomas. Accumulations of bacilli within macrophages, visible as basophilic material in the cytoplasm on H&E stain, are called globi. Fite stain reveals numerous organisms (Fig. 10.4).

Differential Diagnosis

To differentiate sarcoidosis from tuberculoid leprosy with a negative Fite stain, an S-100 protein stain may demonstrate a nerve in the center of the granulomas and thus favor tuberculoid leprosy.

Tertiary Syphilis

Clinical

Generalized disease with skin/mucosal lesions and cardiovascular and neurological manifestations. Occurs years after initial infection. Two types of skin lesions. Superficial nodular lesions are red-brown scaly nodules with a serpiginous advancing border. Gummatous lesions are subcutaneous swellings that ulcerate.

Histology

Marked endothelial swelling in both forms. Gummas have granulomas with large areas of acellular (gummatous) necrosis. The surrounding chronic inflammation includes plasma cells. Nodular lesions show hyperkeratosis overlying an atrophic

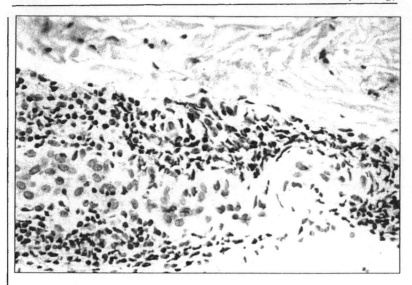

Fig 10.3. **Tuberculoid leprosy**: Linear non-necrotizing granulomas along nerve tracks. Fite stain is often negative in this type of leprosy.

epidermis and a superficial and deep mixed inflammatory infiltrate containing plasma cells and granulomas.

Stains
Silver stains for spirochetes are usually negative.

Other Non-Nectrotizing Granulomas
Beryllium granuloma is of historical interest. It occurred when beryllium compounds were used in fluorescent light tubes (up to 1949). Presented as a nonhealing, tender ulcer at site of laceration by broken fluorescent tubes or as systemic disease with pulmonary involvement due to inhalation. Cutaneous lesions of systemic berylliosis show sarcoid-like granulomas. Cutaneous lesions from lacerations show acanthosis and ulceration with sarcoid-like granulomas and prominent eosinophilic, hyalinized necrosis of the dermis. Polarization of the lesions is negative.

Zirconium granuloma results from zirconium compounds in antiperspirant deodorants. Presents as soft, red-brown papules. Thought to be an allergic sensitization. Histology is that of sarcoidosis. Polarization is negative due to small size of zirconium particles.

Cheilitis granulomatosa (Miescher-Melkersson-Rosenthal syndrome) may present with some or all of a classic triad: recurrent labial edema, relapsing facial paralysis, and fissured (scrotal) tongue. Monosymptomatic labial swelling may define syndrome. Histology shows dermal/submucosal edema with scattered granulomas and an infiltrate of plasma cells and lymphocytes. Granulomas may be absent. Need to rule out cutaneous Crohn's disease.

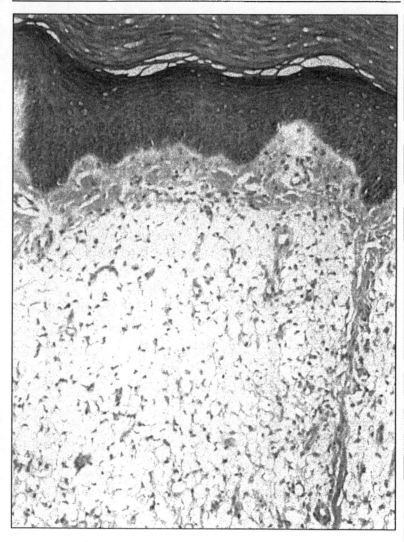

Fig 10.4. **Lepromatous leprosy**: The dermis is completely infiltrated by clear, foamy macrophages which are loaded with bacilli with Fite stain.

Crohn's disease may involve intraoral (pyostomatitis vegetans), perioral, and perianal skin. Occurs in 10-20% of patients. Skin lesions may precede bowel symptoms. Erythema nodosum is often seen associated with Crohn's disease. "Metastatic" Crohn's lesions are seen anywhere in the skin. Histology shows scattered non-necrotizing granulomas in dermis and subcutis with lymphohistiocytic infiltrate. Differential diagnosis includes orificial tuberculosis (will have numerous acid-fast bacilli), sarcoidosis, cheilitis granulomatosa, and rosacea.

Necrotizing Granulomas

Tuberculosis

Clinical

Caused by infection with *Mycobacterium tuberculosis*. Several variants of cutaneous disease will be discussed here. Some variants typically have negative Fite stains. In these cases, correlating the histology with the clinical history and physical examination should favor the correct diagnosis. Stains for other infectious agents (PAS, gram), cultures, and use of PCR for mycobacterial DNA can also be utilized.

Primary (inoculation) tuberculosis from penetrating injury (tattooing, prosector, etc.). Red indurated papule after 1-3 weeks, subsequently ulcerates (tuberculoid chancre). Histology shows suppurative inflammation of dermis initially (1-14 days), then epithelioid cells with granuloma formation (1-6 weeks). Numerous bacilli seen early, few bacilli once granulomas develop.

Lupus vulgaris is most common form of reactivation tuberculosis. Affects primarily young adults, head and neck region. Histology shows variable epidermal changes (atrophic to hyperplastic) and well-demarcated granulomas with variable caseous necrosis and peripheral lymphocytes in the superficial dermis. Moderate numbers of Langhans giant cells in granulomas. Acid-fast bacilli difficult to demonstrate. Can be difficult to distinguish from sarcoidosis (Fig. 10.5).

Tuberculosis verrucosa cutis is uncommon. Results from inoculation of *M. tuberculosis* in a sensitized person. Typically, verrucous plaque on dorsal hand or fingers ("prosector's wart"). Histology shows epidermal hyperplasia with hyperkeratosis and papillomatosis, and necrotizing granulomas in mid-dermis. Acid-fast bacilli may be hard to find.

Scrofuloderma results from tuberculous lymphadenitis extending to skin. Typically presents as draining sinus tract over cervical lymph nodes. Histology shows dermal abscess and/or necrosis with peripheral granulomas. Acid-fast bacilli may be seen on smears, but not always seen in tissue.

Orificial tuberculosis is rare. Results from autoinoculation in patients with advanced pulmonary tuberculosis and causes shallow ulcers at mucocutaneous junctions. Histology shows ulceration and underlying necrotic granulomas with numerous acid-fast bacilli.

Miliary tuberculosis occurs in infants and is rare in immunocompetent persons. Widespread internal involvement (hematogenous spread) with generalized erythematous papules and pustules (2-5 mm). Histology shows dermal microabscess with some peripheral granuloma formation. Numerous acid-fast bacilli within abscess.

Tuberculids

Clinical

Tuberculids are heterogeneous skin lesions in patients with tuberculosis infection elsewhere (lymph nodes most common). It occurs in patients with high degree of immunity and sensitivity to *M. tuberculosis*. No organisms are a

Table 10.2. Some atypical mycobacterial diseases

M. *marinum*—fish tank granuloma
M. *ulcerans*—Buruli ulcer
M. *avium intracellulare*—seen in AIDS
M. *cheloni and* M. *fortuitum* (rapid growers)—iatrogenic with contaminated needle/cannula

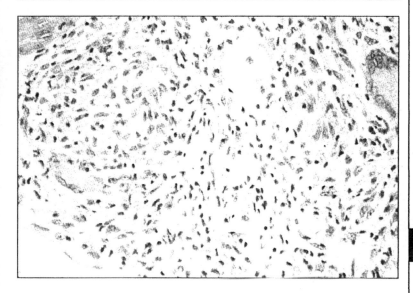

10

Fig 10.5. **Tuberculosis**: Non-necrotizing granulomas with Langhan's type giant cells (probably miliary tuberculosis).

present in tissue or culture, but have been identified by PCR techniques. Three tuberculid diseases are recognized.

Erythema induratum (Bazin's disease, nodular vasculitis) is a panniculitis. Presents with blue-red plaques and nodules on lower legs (calves) which tend to ulcerate. Females are more frequently involved than males. Recurrences may be precipitated by cold weather. Histology shows lobular panniculitis with extensive fat necrosis and granuloma formation. Granulomas may be absent in up to 1/3 of cases. Nomenclature: "erythema induratum" if associated with tuberculosis, "nodular vasculitis" if not.

Lichen scrofulosorum presents in children and young adults with yellow or brown scaly follicular papules (0.5B3mm). Lesions heal without scarring. Histology shows superficial dermal granulomas, generally without necrosis, around hair follicles or sweat ducts.

Papulonecrotic tuberculid presents as erythematous papules on the extremities in a symmetrical distribution. Lesions ulcerate and heal with depressed scars. Histology shows leukocytoclastic vasculitis in early lesions. As lesions evolve, there is ulceration and a wedge-shaped area of dermal necrosis (broad base towards epidermis) with a peripheral palisade of histiocytes and lymphocytes and only occasional well-formed granulomas.

Atypical mycobacterium includes a miscellaneous group of nontuberculous mycobacteria. *M. marinum* is the most common. Produces a variety of lesions and the histologic changes are not species-specific. Atypical mycobacteria should be considered when there are poorly-formed granulomas with chronic inflammation and foci of acute inflammation and necrosis.

Other Necrotizing Granulomas

Leprosy can have necrotizing granulomas. See above discussion for details.

Deep fungal infections are also associated with necrotizing granulomas with mixed cellularity. The specific entities are discussed in detail in Chapter 14. A PAS or GMS stain is obtained on all necrotizing granulomas to rule out fungus.

Foreign Body Granulomas

Foreign body granulomas may be caused by exogenous (foreign) material or endogenous material that is either misplaced or altered such that it is recognized as foreign. The histology is similar for all of these entities. There are histiocytes, giant cells, plasma cells, lymphocytes and neutrophils around the foreign body. The giant cells are usually of the foreign body type, but Langhans giant cells are often present as well. Polarization may be useful in identifying the foreign body.

Foreign Body Granulomas: Exogenous

Paraffinomas follow injections of oily substances (paraffin). Present with irregular, plaque-like indurations of skin and subcutis. Histology shows "Swiss cheese" appearance with round to ovoid cavities surrounded by fibrosis and granulomatous inflammation.

Silicon granulomas follow injection of silicone for cosmetic purposes or leakage from breast implants. Presents as subcutaneous nodules and plaques. Histology is similar to paraffinoma with "Swiss cheese" appearance. Some residual silicone may be seen after processing as colorless, irregularly shaped, refractile, nonpolarizable material.

Tattoo reactions usually occur with red dyes (cinnabar). Histologically, may be granulomatous (foreign body type or sarcoidal) or nongranulomatous (with lichenoid response and pigment-containing macrophages).

Starch/Talc granulomas result from talc or corn starch present on surgical gloves. Talc crystals are needle-shaped and have white birefringence with polarization. Starch granules react with PAS and GMS stains and have birefringence (with classic Maltese cross configuration) with polarization.

Other granulomas include cactus (spicules are PAS-positive) and sea urchin granulomas (may have doubly refractile material on polarization due to silica in urchin spines).

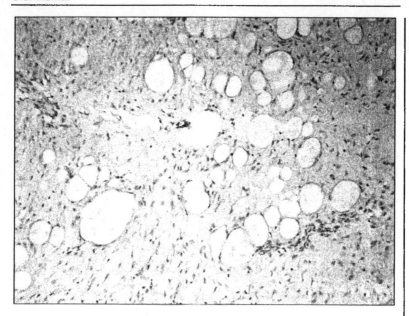

Fig 10.6. **Paraffinoma**: The dermis shows fibrosis and irregular, cystic spaces which appear clear. The paraffin or other lipid material has been dissolved during processing.

10

Foreign Body Granulomas: Endogenous

Gout

Clinical
Early stages have recurring bouts of acute arthritis. Later stages have deposits of monosodium urate involving joints. In late stage, can have cutaneous deposits of urate (tophi). Tophus lesions present on helix, elbows, and fingers and toes. Tophi can become large (several centimeters) and extrude chalky material.

Histology
Fixation in alcohol may preserve the urate crystals, which are largely dissolved in aqueous fixatives such as formalin. The epidermis may be ulcerated. There are variably sized deposits of amorphous, amphophilic material with parallel, needle-shaped clefts in dermis and subcutis. The deposits are surrounded by a granulomatous reaction with many foreign body giant cells. In alcohol-fixed specimens, the crystals are brown and doubly refractile with polarized light (Fig. 10.7).

Keratin Granulomas
Keratin granulomas are very common. They are often seen with ruptured epidermal inclusion cysts, but are also present in other disorders: ruptured follicles/

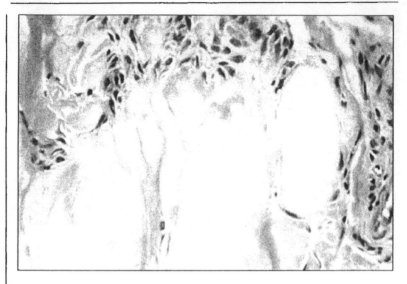

Fig 10.7. **Gout**: Needle-shaped crystals and amorphous, birefringent material surrounded by a giant cell foreign body granulomatous inflammation.

follicular cysts and ruptured horn cysts in trichoepithelioma, for example. The granulomas are usually adjacent to the cyst and contain numerous foreign body giant cells. If no cyst is identified, keratin debris may be noted with the granuloma as slivers of anucleate eosinophilic debris.

Necrobiotic Granulomas

Necrobiotic granuloma refers to dermal granulomas that have altered dermal collagen in the center with the granulomatous inflammation often forming a "palisade" around it. The "necrobiotic" collagen will have loss of definition, decreased nuclei, and altered staining (eosinophilia or basophilia). It is a distinct form of granuloma seen in the following diseases.

Granuloma Annulare

Clinical

Usually presents in children and young adults with erythematous to skin-colored papules that group to form annular plaques. Affects hands, feet, arms, and legs. Females are more frequently affected than males. Clinical variants include localized, generalized, perforating, and subcutaneous (deep). Most lesions will resolve spontaneously, but may recur.

Histology

There are necrobiotic granulomas in the upper to mid-dermis (extend to subcutis in subcutaneous variant). Incomplete (interstitial) granuloma annulare may not have prominent necrobiosis. There is usually increased acid mucopolysaccharide associated with the granulomas. An alcian blue or colloidal iron stain can help identify the increased mucin. The presence of histiocytic cells in the dermis is always suggestive of granuloma annulare (Fig. 10.8).

Necrobiosis Lipoidica Diabeticorum

Clinical
Many cases are associated with diabetes, but occurs in nondiabetics as well. Presents with irregularly demarcated patches or plaques that have an atrophic, yellow-brown center and a raised red edge. Occurs predominantly on legs (shins) in children and young adults. Females are more commonly affected than males.

Histology
Characteristically, the full thickness of the dermis (and even subcutis) is involved and several layers of necrobiosis can be seen (sandwich changes). The changes are more marked in the lower two-thirds. The inflammatory infiltrate may include plasma cells. There is epidermal atrophy and telangiectasia in the compressed papillary dermis (Fig. 10.9).

Differential Diagnosis
Granuloma annulare and necrobiosis lipoidica can be difficult to distinguish. Superficial location and increased mucin in granulomas favors granuloma annulare. Epidermal atrophy, extensive dermal involvement, and plasma cell infiltrate favors necrobiosis lipoidica.

Rheumatoid Nodule

Clinical
Occurs in about 20% of patients with rheumatoid arthritis. It presents over extensor surfaces near joints. Vary from millimeters to several centimeters in size. May be solitary or multiple. Associated with high titer rheumatoid factor. Have also been reported with systemic lupus erythematosus.

Histology
Nodules are present in the deep dermis/subcutis and contain large, well-demarcated areas of necrobiotic collagen surrounded by a "palisade" of lymphocytes and histiocytes, including foreign body giant cells. The degenerated collagen appears as eosinophilic, amorphous material. Mucin is minimal or absent.

Differential Diagnosis
The lack of mucin and presence of foreign-body giant cells favors rheumatoid nodule over subcutaneous granuloma annulare.

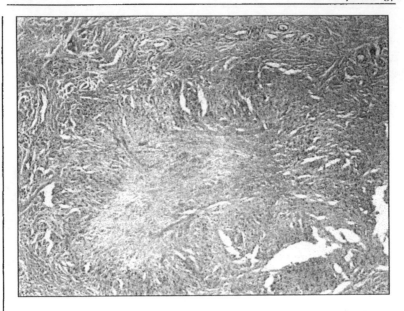

Fig 10.8. **Granuloma annulare**: Irregularly-shaped necrobiotic area in the dermis, surrounded by macrophages with pallisading. Mucin deposition is characteristic.

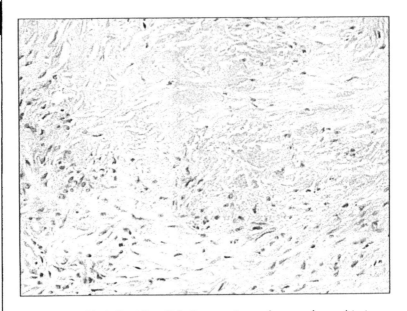

Fig 10.9. **Necrobiosis lipoidica diabeticorum**: Layered areas of necrobiosis surrounded by macrophages (sandwich sign). Giant cells are commonly seen.

Annular Elastolytic Granuloma (Actinic Granuloma)

Clinical

Occurs on sun-exposed skin. Patients are usually more than 40 years old. Presents like granuloma annulare with large, annular plaques. May represent granuloma annulare on sun-exposed skin. The best place to biopsy is near the periphery of the lesion, such that the central area, annular ring, and perilesional skin are sampled in the biopsy.

Histology

The changes are in the dermis. An optimal biopsy will show the central area to have decreased elastic fibers with normal or thickened collagen. The annular ring contains numerous histiocytes and foreign body giant cells, often containing phagocytized elastic fibers. Asteroid bodies may be present (and stain like elastic fibers). The perilesional zone contains increased elastotic material. Mucin is absent. A VVG stain for elastic fibers can contrast the three zones.

Differential Diagnosis

Granuloma annulare is the main differential diagnosis. The absence of mucin, presence of multiple nuclei in the foreign body giant cells (up to 12), and the characteristic absence of elastotic material in the center of the lesion favor annular elastolytic granuloma.

Chondrodermatitis Nodularis Helices

10

Clinical

Presents as solitary, 4-6 mm, dull, tender nodule on the upper helix in patients over 50. Lesions persist indefinitely, with up to 20% recurrence after cautery and curettage. Intermittent crusting may mimic basal cell carcinoma.

Histology

Characteristically, epidermal ulceration with edematous and acanthotic epithelial margins is present. The dermis shows necrobiotic collagen surrounded by granulation tissue. Histiocytes may palisade around the necrobiosis. The underlying cartilage shows degenerative changes with altered staining and hyalinization. Calcification and ossification may occur (Fig. 10.10).

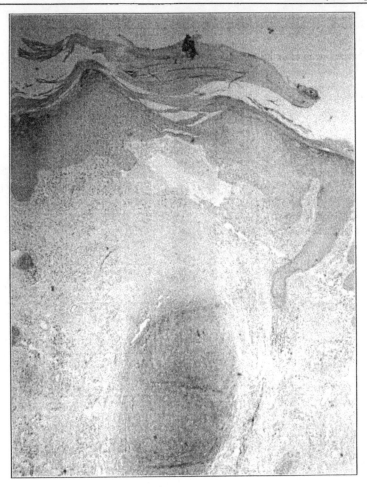

Fig 10.10. **Chondrodermatitis nodularis helices**: The cartilage is focally degenerated. The dermis above is necrobiotic and is extruded through the epidermis.

CHAPTER 11

Dermal Necrosis and Transepidermal Elimination

San-Hwan Chen

Section 26. Necrosis of the Dermis Without Granuloma

Necrosis implies cell death. If the necrosed tissues are disposed of, an ulcer ensues. Morphologically necrosis has been classified as liquefactive (most ulcers, wet gangrene), ischemic (infarcts, tissues are mummified-ghost cells), caseous (specific for tuberculosis), and necrobiotic (incomplete necrosis). Apoptosis refers to programmed cell death, and generally refers to normal tissue turnover. According to etiology, ulcers can be traumatic, inflammatory, ischemic, and infectious. Often more than one mechanism operates in producing an ulcer, for instance after trauma there is an inflammatory response that can enhance the death of some tissues. The newly formed ulcer can then become infected, which triggers more inflammation and cell death. All these processes can be compounded if there is poor circulation, as is often the case in leg ulcers of elderly patients.

Necrosis is often one component of granulomatous inflammation, as seen in the preceding chapter. In this section we refer to purely inflammatory ulcers, inflammatory ulcers with an infectious etiologic agent, and ischemic ulcers with little or no inflammation.

Ulcers

Clinical

An ulcer results from the destruction of the epidermis and at least the upper dermis, and presents as an eroded area in the skin. Ulcerations of the skin have many etiologies, including trauma, infection, and tissue infarction due to occlusion or constriction of blood vessels. Leg ulcers are common problems, mainly associated with disease of the veins (80-90%) and arteries (5-10%). Venous ulcers tend to occur around the malleoli and frequently have irregular bases of granulation tissue or fibrinous material (Fig. 11.1). Arterial ulcers are frequently located at sites of pressure or trauma and are usually extremely painful. They frequently have a punched-out appearance with minimal granulation tissue.

Histology

A chronic ulcer is characterized by central suppurative necrosis with neutrophils adjacent to the necrosis and with granulation tissue, fibrosis, and reactive epithelium at the periphery.

Dermatopathology, edited by Ramón L. Sánchez and Sharon S. Raimer. ©2001 Landes Bioscience.

Table 11.1. *Inflammatory ulcers*

Ulcers
Pyoderma gangrenosum
Sweet's syndrome
Vascular ulcer/atrophic blanche
Ant, Spider bites

Fig 11.1. **Venous ulcer**: Ulceration of the epidermis and dermis with acute and chronic inflammation. The capillaries show fibrin deposits.

Ant/Spider bites

Clinical

Fire ant bites classically result in the formation of small pustules on erythematous bases. Early lesions show severe edema. Intraepidermal pustules filled with neutrophils and eosinophils are present within 72 hours. The dermis contains a dense mixed cellular infiltrate with basophilic necrosis (Fig. 11.2).

The brown recluse spider and the black widow spider are significant causes of disease in the United States. Seventy five percent of patients develop painful necrotic cutaneous lesions after brown recluse bites. Black widow spider bites begin as painless bites, followed by swelling and tenderness at bite sites and painful regional lymphadenopathy, then systemic symptoms develop (headache, backache, and abdomi-

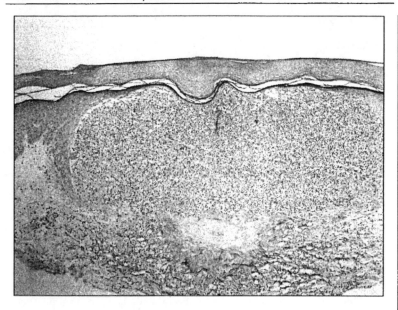

Fig 11.2. **Ant bite**: The upper dermis shows a large pustule. The dermis underneath shows basophilic necrosis admixed with an acute infiltrate.

nal pain). The bites of other spiders may also result in the formation of erythematous papules, nodules, or plaques which ulcerate.

Histology
There is necrosis of the epidermis and dermis with eschar formation. The area of necrosis is surrounded by a neutrophilic infiltrate admixed with lymphocytes and plasma cells.

Infectious Ulcers

Anthrax

Clinical
Caused by *Bacillus anthracis*, a large gram-positive encapsulated bacillus, occurring occasionally among workers in tanneries and wool-scouring mills. The lesions start as papules which coalesce into large hemorrhagic pustules, then rupture and form a thick black eschar on the top surrounded by marked erythema and edema.

Histology
Histology shows epidermal necrosis and vasculitis. Hemorrhage and variable acute and chronic inflammatory infiltrates are observed in the dermis. *Anthrax bacilli* are present in large numbers when stained with Gram stain.

11

Chancroid

Clinical

Caused by *Hemophilus ducreyi*, a Gram-negative coccobacillus. It is sexually transmitted leading to a genital ulcer with an undermined border and associated with inguinal lymphadenitis.

Histology

Histologically three zones are present: the superficial zone containing a narrow band of neutrophils, fibrin, erythrocytes and necrotic tissue; the next zone containing a wide area of granulation tissue; and the deeper zone composed of dense infiltrate of plasma cells and lymphoid cells. Bacilli are present in the superficial zone, lying in parallel chains.

Tularemia

Clinical

Caused by *Francisella tularensis*, and is usually transmitted through direct contact with rodents. It often occurs in small epidemics. Six clinical patterns depending on the route of exposure have been described. **Ulceroglandular** is most common (80% of the cases). Presents as one or several painful ulcers at the site of infection, usually on the hands. Draining subcutaneous nodules may form along lymphatic vessels accompanied by tender regional lymphadenopathy and constitutional symptoms. The lesions heal in two to five weeks.

Histology

The histology of a primary ulcer shows a nonspecific inflammatory infiltrate associated with a granulomatous reaction with central necrosis. Subcutaneous nodules show multiple granulomas deep in the dermis and extending into subcutaneous tissue with extensive central necrosis.

Furuncle/Carbuncle

Clinical

A furuncle or boil is a staphylococcal infection of the hair follicle. It presents as a red painful nodule containing pus, which drains spontaneously. A carbuncle is a staphylococcal infection involving several adjacent follicles.

Histology

Histology shows perifollicular necrosis with neutrophils and fibrinoid material. Generally an abscess occurs at the deep end of the lesion. Small clusters of staphylococci can be seen in the abscess with Gram stain.

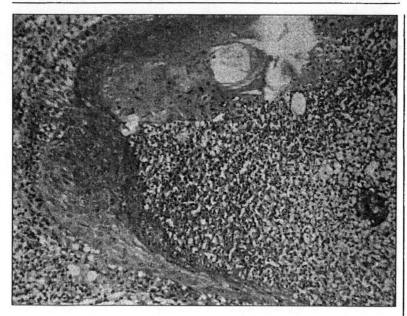

Fig 11.3. **Furuncle**: A large hair follicle associated abscess. Coalescence of several follicles is common.

Noninflammatory Ulcers

Infarct

Clinical

A cutaneous infarct is characterized by ischemic necrosis of the dermis resulting from occlusion of a blood vessel. Such occlusion may have different etiologies such as vasculitis, bacterial embolism and atheroembolism. Lesions appear stellate in shape, and are often surrounded by livedo reticularis. They begin as purpuric macules and papules, then become gray or black in color and develop ulceration. The lesions occur most commonly on distal extremities.

Section 27. Transepidermal Elimination

Perforating Disorders

The perforating disorders are a group of papulo-nodular skin diseases characterized by hyperkeratotic plugs in which dermal connective tissue is eliminated or perforates through the epidermis. There are two perforating diseases which appear to be genetic in origin. They are reactive perforating collagenosis, in which collagen fibers perforate the epidermis and elastosis perforans serpiginosa, in which the perforating material is composed primarily of elastic fibers. Acquired perforating

---

Table 11.2. Conditions with transepithelial elimination

Reactive perforating collagenosis
Elastosis perforans serpiginosa
Acquired perforating dermatosis
Perforating folliculitis
Kyrle's disease
Perforating granuloma annulare
Chondrodermatitis nodularis helices
Tumoral transepithelial elimination

---

dermatosis (uremic follicular hyperkeratosis) is usually acquired in adulthood and is associated frequently with diabetes mellitus or renal failure. Perforating folliculitis has been considered a perforating disease by many authors. This is probably not justified, as this disorder does not seem to be a specific entity. Rupture of follicles can occur for a wide variety of causes, such as acne, physical trauma, or infarction of hair follicles (Fig. 11.4).

Elastosis Perforans Serpiginosa

Clinical
Occurs in the second decade with a male predominance (4:1). Flesh colored or red keratotic papules with adherent plugs in a serpiginous distribution or occasionally randomly distributed, with the back and side of the neck often being involved. The condition occurs in association with Down's syndrome, Ehler-Danlos syndrome, osteogenesis imperfecta, pseudoxanthoma elasticum, Marfan's syndrome, and rarely as a complication of penicillamine therapy.

Histology
Histology shows transepidermal elimination of abnormal elastic tissue with marked increase in both the amount and the size of the elastic fibers in the upper dermis.

Reactive Perforating Collagenosis

Clinical
A very rare inherited disorder with an unusual skin reaction to mild trauma. Damaged collagen is extruded though the epidermis. Hyperkeratotic papules 5-8 mm in diameter develop over 3-4 weeks following superficial trauma. Koebnerization frequently occurs.

Histology
Histology shows a cup-shaped channel in the epidermis containing parakeratotic keratin, basophilic collagen and picnotic nuclei of inflammatory cells. Collagen fibers may be seen perforating through the epidermis.

Stain
VVG (Verhoff van Gieson) stain is helpful because collagen fibers stain red and elastic fibers stain black.

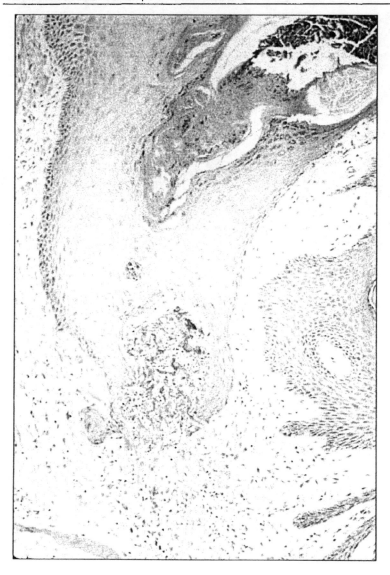

Fig 11.4. **Perforating folliculitis**: A hair follicle appears distorted and plugged, show-ing an area of inflammation and necrosis which is in continuity with the dermis.

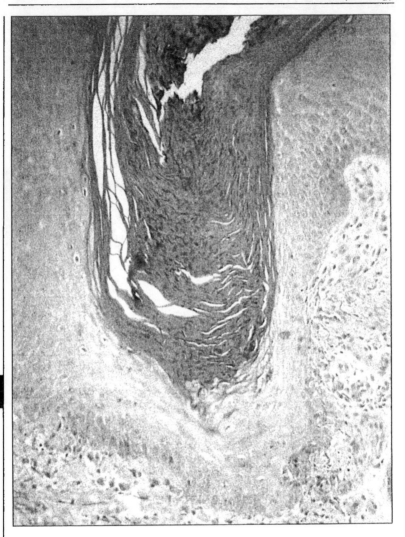

Fig 11.5. **Uremic follicular hyperkeratosis**: An epidermal, keratin-filled plug with crust, and occasional communication with the dermis which appears focally inflamed.

Acquired Perforating Dermatosis

Clinical

Uremic follicular hyperkeratosis is a term used for perforating disease arising in adulthood, usually in association with diabetes or renal failure, but occasionally secondary to internal malignancy or liver disease. It has been estimated that 5-10% of patients undergoing hemodialysis develop this disorder. Acquired perforating dermatosis most commonly presents as hyperkeratotic papules or nodules on the legs, but widespread lesions are not uncommon. The Koebner phenomenon may occur.

Histology

Findings in acquired perforating dermatitis vary according to the stage of the lesion biopsied. The histology may be identical to reactive perforating collagenosis, elastosis perforans serpiginosa, or perforating folliculitis, or it may be less specific with amorphous degenerated material within the perforations (Fig. 11.5).

Kyrle, in 1916, described an acquired perforating disease that he called **hyperkeratosis follicularis et parafollicularis in cutem penetrans**, and it is reasonable to use Kyrle's disease as a synonym for acquired perforating dermatosis. Unfortunately the literature concerning Kyrle's disease is quite confusing, as many authors have used different criteria to define the entity.

Other Disorders with Transepidermal Elimination

This includes a variety of unrelated disorders in which transepidermal elimination of a substance occurs. The epidermis often becomes hyperplastic and eventually surrounds the material and causes its elimination by normal keratinocyte maturation. Some of these disorders include the perforation of foreign material (such as calcium, silica, or wood splinters), infectious agents (such as chromomycosis, botryomycosis, or spirochetes), granulomas (such as perforating granuloma annulare, necrobiosis lipoidica, or sarcoidosis), neoplastic cells, also known as tumoral transepidermal elimination (occurring with malignancies such as melanoma, basal cell carcinoma, Paget's disease, or mycosis fungoides), or other degenerated endogenous substances (such as chondrodermatitis nodularis helices, hematoma, perforating pseudoxanthoma elasticum, and perforating periumbilical calcific elastosis).

11

CHAPTER 12

Purpuras, Vasculitis, Emboli and Thrombi

Ramón L. Sánchez and Sharon S. Raimer

Section 28. Purpura

Purpura

Purpura is a clinical term which implies hemorrhage into the dermis, i.e., extravasation of red blood cells. The hemorrhage in the dermis is clinically subclassified according to size into petechiae with a size less than 3 mm in diameter vs ecchymoses which are larger hemorrhages. Collections of blood in the skin or subcutaneous tissue are defined as hematomas. In general, purpuras imply damage to the blood vessel either traumatic, obstructive or inflammatory. Histopathologically we classify purpuras into noninflammatory or inflammatory. The former are usually associated with obstruction of capillaries or arterioles by fibrin thrombi, platelets, immunoglobulin complexes, bacterial thrombi, cholesterol and calcium deposits, or other causes. Inflammatory purpura usually imply vasculitis.

Histopathology of noninflammatory purpura shows eosinophilic thrombi of vessels of mid and superficial dermis associated with hemorrhage and only mild, if any, inflammation. There is often a variable degree of necrosis of the dermis and/or epidermis. The etiology of the process should be pursued, usually clinically, although occasionally the biopsy can pinpoint the cause.

Non-Inflammatory Purpuras

Thrombotic Thrombocytopenic Purpura

Clinical

Thrombotic thrombocytopenic purpura is clinically characterized by hemolytic anemia, thrombocytopenia, neurological symptoms, renal disease and fever.

Histology

There is fibrin thrombi present in the blood vessels of the dermis.

Purpura Fulminans/Disseminated Intravascular Coagulation

Clinical

There is activation of the coagulation system with formation of fibrin thrombi and also hemorrhage. Usually it follows infections, particularly meningococcemia, neoplasms, a side effect of some obstetric procedures, burns, liver disease and snake bites. Clinically the purpura and hemorrhage is widespread.

Dermatopathology, edited by Ramón L. Sánchez and Sharon S. Raimer. ©2001 Landes Bioscience.

Table 12.1. Non-inflammatory purpuras

ITP, TTP
Purpura fulminans/DIC
Cryoglobulinemias/Cryofibrinogenemias
Anticoagulant necrosis
Lupus anticoagulant/antiphospholipid antibody
Calciphilaxis
Atrophie blanche
Cholesterol embolism
Senile purpura
Scurvy
Sneddon's syndrome

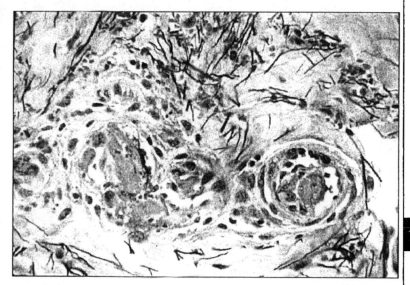

12

Fig 12.1. **Cryofibrinogenemias**: The capillaries in the dermis are filled with fibrin thrombi. No significant inflammation is seen in the dermis. Movat stain.

Histology
Early lesions show fibrin thrombi in capillaries and venules.

Cryoglobulinemias

Clinical
Cryoglobulinemias are immunoglobulins that precipitate with the lowering of temperature. They can be monoclonal or mixed. The former are usually IgG or IgM cryoglobulins (type 1) and usually are seen associated with myeloma, Waldenstrom's macroglobulinemia and chronic lymphocytic leukemia. Mixed cryoglobulinemias are at least in part polyclonal (types 2 and 3), usually form immune complexes and are

seen in rheumatoid arthritis, SLE, Sjogren's syndrome, hepatitis C or B, and infections by Epstein-Barr virus. A similar clinical and histologic presentation is sometimes seen as a result of cryofibrinogenemia, either idiopathic or associated with an infectious or neoplastic condition (Fig. 12.1).

Histology

Monoclonal cryoglobulinemia shows the typical findings of noninflammatory purpuras while mixed cryoglobulinemias exhibit the changes of a leukocytoclastic vasculitis.

Anticoagulant Necrosis

Clinical

Anticoagulant necrosis is usually seen on patients on Warfarin, a Coumarin derivative, although it has been also described after treatment with heparin. Warfarin necrosis is related to low levels of protein C, a vitamin K dependent plasma protein.

Histology

In addition to fibrin thrombi there is extensive hemorrhage and infarction of the dermis and epidermis.

Lupus Anticoagulant/Antiphospholipid Antibody

Clinical

Lupus anticoagulant or anticardiolipin antibodies are seen in 20-50% of patients with lupus erythematosus. Some of these patients also have the antiphospholipid syndrome which is characterized by antibodies or autoantibodies directed against phospholipids and associated with episodes of thrombosis, fetal loss and thrombocytopenia. Clinical findings include livedo reticularis, thrombophlebitis, and sometime even infarction of gangrene.

Histology

The findings are similar to other noninflammatory purpuras.

Calciphylaxis

Clinical

Calciphylaxis is characterized by large nonresolving ulcers on abdomen and thighs and legs, in patients with renal failure in which there is usually an elevated calcium-phosphate product and secondary hyperparathyroidism.

Histology

Calcium deposits are seen in the wall and sometimes in the lumen of the small arteries in the subcutaneous tissue as well as deep dermis. The calcification along blood vessels is sometimes so prominent that it can be visualized by simple x-rays.

Other related purpuras include **senile purpura**, usually associated with atrophy of the epidermis and dermis as well as solar elastosis in other patients with easy

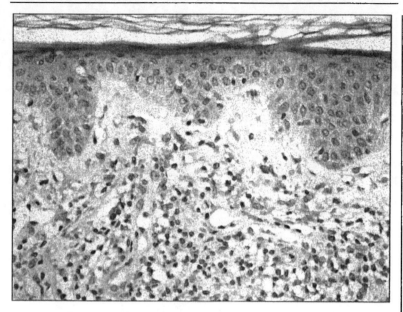

Fig 12.2. **Purpura pigmentosa chronica**: There is a lymphocytic infiltrate surrounding the superficial capillary plexus with extravasation of red blood cells.

bruise ability. **Scurvy** associated with a deficiency of vitamin C (ascorbic acid) is characterized by perifollicular hemorrhage, abnormal hair growth, bleeding gums and poor wound healing. Histologically there are extravasated blood cells around blood vessels and often in a perifollicular distribution. Hemosiderin deposits are also seen. Sneddon's syndrome is characterized by livedo reticularis and ischemic cerebrovascular manifestations. Histologically the changes are minimal or nonspecific, although occasionally the small blood vessels of the dermis show fibrin thrombi.

Inflammatory Purpuras

Capillaritis

The inflammatory purpuras are usually associated with vasculitis (see section 31). Inflammatory purpuras without vasculitis are the so-called capillarites, and extravasation of red blood cells in the dermis or epidermis associated with an inflammatory dermatosis is seen in conditions such as pleva, pityriasis rosea, lupus erythematosus, and syphilis.

Capillaritis/Purpura Pigmentosa

Clinical

Capillaritis/Purpura pigmentosa clinically is characterized by petechiae or hemorrhage often in the lower part of the legs, usually long standing. According to the clinical and histologic appearance, the purpura pigmentosas have been classified as

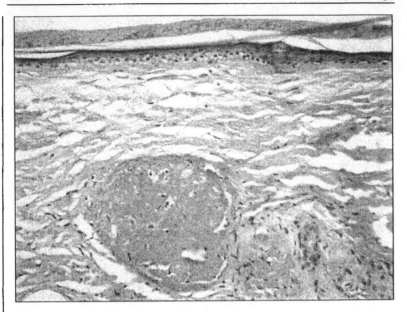

Fig 12.3. **Atrophy blanche**: The epidermis is atrophic. The dermis shows fibrin deposition in the wall of the capillaries, some of which are also thrombosed.

purpura annularis telangiectodes of Majocchi, progressive pigmentary dermatosis of Schamberg, pigmented purpuric lichenoid dermatitis of Gougerot and Blum, and eczematoid-like purpura of Doucas and Kapetanakis. A localized variant is lichen aureus in which patches of flat papules with a rust, copper, or orange color are seen.

Histology

The purpura pigmentosa chronica or capillaritis is characterized by a superficial perivascular lymphocytic infiltrate associated with extravasation of red blood cells (Fig. 12.2). The infiltrate sometimes involves the basal layer with formation of hydropic degeneration. The main differential diagnosis, both clinically and histopathologically is with vasculitis. The capillarites lack neutrophils and nuclear dust, which are present in vasculitis even in the earlier changes.

Section 29. Fibrin Deposition and Emboli

Fibrin thrombi are seen in many of the purpuras, either noninflammatory or inflammatory. Fibrin deposition is seen also in the wall of capillaries in atrophy blanche where it is a diagnostic feature. Fibrin deposition is also present in the wall and around the wall of capillaries, arterioles and postcapillary venules in leukocytoclastic vasculitis (see section 31).

Atrophy blanche is clinically characterized by purpuric papules and plaques that eventually become ulcers, particularly in the lower part of the legs or ankles.

The ulcers heal after long periods of time with atrophic white scars. Histopathologically, the blood vessels in the dermis show hyalinization of the wall and occasionally fibrin deposition and/or thrombi (Fig. 12.3).

Cholesterol embolism is characterized by petechiae, gangrene and ulceration of digits particularly on the feet, the result of embolism from atherosclerotic plaques. Histopathologically, clefts from cholesterol crystals are seen in arteries from the dermis. A fibrin thrombus is usually associated with the embolic material.

Section 30. Organized Thrombi and Vascular Clot

Thrombi develop when the coagulation factors in the blood and the vessel wall are released by hemorrhage or injury to the endothelium of the blood vessel. Interrupted or slowed blood flow such as in embolism also can start the coagulation cascade. A blood clot is composed of fibrin, white and red blood cells. Depending on the size of the clot, layering of the white cells and platelets occur, producing characteristic concentric rings. With time clots become organized with introduction of granulation tissue, capillary formations, fibroblasts and inflammatory cells. There is an attempt to recannalize the thrombus.

Thrombophlebitis is clinically seen after injury of a superficial vein, for instance in the cubital fossa after venipuncture or in the veins of the legs sometimes associated with trauma and/or varicose veins. Mondor's thrombophlebitis refers to a painful cord between the breast and the axilla. The skin overlying the involved vessel is red, warm and painful. A hard cord-like structure that corresponds to the thrombosed vessel can often be palpated. Histologically the wall of the blood vessel is infiltrated by neutrophils. The lumen is dilated and filled by a clot admixed with numerous white blood cells including neutrophils and some lymphocytes (Fig. 12.6).

Thrombosed Hemangioma

Clinical
It is not uncommon for hemangiomas to become partially thrombosed. In that situation there is often sudden onset of redness and tumor formation over a pre-existing lesion.

Histology
A clot or partially organized thrombus is seen filling up a dilated vessel. Adjacent to this area other vascular lumen characteristic of hemangioma are seen.

Superficial Migratory Thrombophlebitis (AguadePinol)

Clinical
Clinically it presents as painful nodules usually on legs or thighs, deep seated, similar to the presentation of paniculitis. With time the nodules move in location slightly.

12

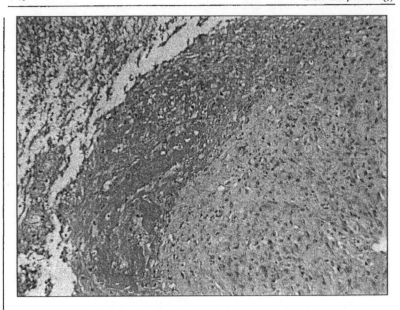

Fig 12.4. **Thrombophlebitis**: The lumen of a large vein is filled by a fibrin thrombus with necrotic debri. The wall of the vessel shows acute and chronic inflammation.

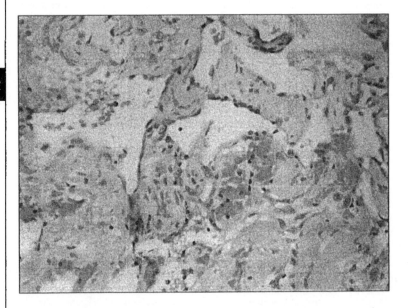

Fig 12.5. **Masson's pseudoangiosarcoma**: a thrombosed lumen has been recannalized showing papillary formations lined by endothelial cells.

Table 12.2. Etiopathologic factors associated with vasculitis

---

Infections:
    Hepatitis B/C
    Viral infections: (E.G., Herpes)
    Bacterial/fungal infections
    Drugs
    Collagen vascular disease
    Lupus erythematosus
    Rheumatoid arthritis
    Sjogren's syndrome
    Cryglobulinemia
    Paraneoplastic vasculitis

Histology

A biopsy shows a dilated vein in the subcutaneous tissue whose lumen is thrombosed and its wall shows a predominantly neutrophilic infiltrate. The fat around it as well as the lower dermis shows a predominantly lobular paniculitis with acute and chronic inflammatory cells.

Masson's Pseudoangiosarcoma (Intravascular Papillary Endothelial Hyperplasia)

Clinical

Clinically there is usually a nodule that may be painful, in any location, particularly head, neck and hands.

Histology

Histologically the remnants of large dilated blood vessels are usually apparent. The blood vessel lumen is filled with a fibrovascular proliferation with multiple papillary projections lined up by plump endothelial cells (Fig. 12.5). Between the papillae irregular vascular spaces which may resemble anastomosing vascular channels are seen. For that reason, the histologic picture can be worrisome and even suggesting the possibility of an angiosarcoma. The fact that all this neovascular proliferation occurs inside of a vessel wall and the finding of typical organized thrombis elsewhere in the specimen, points out to the benign nature of this condition. Atypical intravascular endothelial hyperplasia is a similar process in which the endothelial cells lining papillary projections are markedly atypical. Again, the intravascular nature of the process points to the reactive, benign nature of this entity. Similar lesions with marked endothelial atypia have been described in the lip. A related entity is intravascular pyogenic granuloma, which shows characteristic findings of lobular capillary hemangioma inside a blood vessel lumen.

12

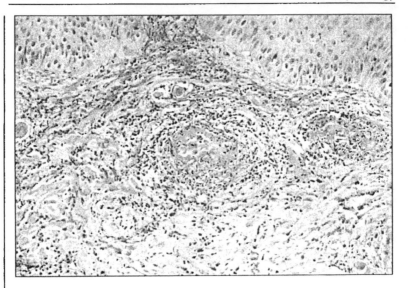

Fig 12.6. **Leukocytoclastic vasculitis**: Fibrinoid necrosis of the vessel walls with thrombosis, neutrophilic infiltrate, leukocytoclasia and extravasation of RBCs.

Section 31. Vasculitis

Vasculitis can be divided roughly into small vessel and large vessel vasculitis. The former is usually superficial while large vessel vasculitis is usually deep and may involve blood vessels in organs other than the skin. In general, neutrophils are required to establish a diagnosis of vasculitis. The neutrophils involve the vessel wall and produce destruction by releasing enzymes. In the process, many of the neutrophils die and the nuclear remnants (karyorrhexis, leukocytoclasia) are seen around the vessel. Other findings exuding from the damage to the vessel wall are endothelial swelling, extravasation of red blood cells, fibrin deposition in the vessel wall and in concentric layers around the vessel, and some degree of necrosis of the dermis or the tissue where the vessel is located (Fig. 12.6). Many instances of leukocytoclastic vasculitis are triggered by immune complex deposition in the vessel wall. Although karyorrhexis is the whole mark of vasculitis, it can be seen also in some infectious processes such as disseminated histoplasmosis in immunocompromised patients and in insect bites. In general, however, the presence of nuclear dust, even a small amount, is indicative of vasculitis. In large vessel vasculitis, either on veins or in arteries the diagnostic feature is usually the presence of neutrophils with nuclear dust in the wall of the blood vessel, which usually appears thrombosed. The inflammatory process extends around the blood vessel and there is usually associated ischemic changes in the neighboring tissue.

Some inflammatory dermatoses are classified as lymphocytic vasculitis, including pityriasis lichenoides et varioliformes acuta or pleva, pyoderma gangrenosum and erythema multiforme. Proponents of this type of vasculitis imply that there is

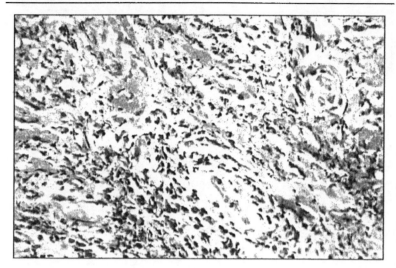

Fig 12.7. **Erythema elevatum diutinum**: Fibrin deposition in the vessel walls and a mixed infiltrate with leukocytoclasia.

destruction of the vessel wall by lymphocytes and the lymphokines secreted. On practical grounds, it is difficult to distinguish a perivascular lymphocytic infiltrate from a lymphocytic vasculitis.

Small Vessel Vasculitis (Leukocytoclastic Vasculitis)(LVC)

Clinical

Clinically it is often referred to as palpable purpura because of the characteristic nonblanchable purpuric macules which are slightly raised over the skin surface. This type of vasculitis is also referred to as necrotizing vasculitis. Typically the purpuric lesions appear first in distal extremities, particularly dorsal feet and legs, as well as hands and forearms and they may extend from there elsewhere. There may be small petechial lesions, or larger ecchymotic macules. Leukocytoclastic vasculitis can be seen at any age, although patients are often older, particularly when there is an association with drug intake. Many etiologic factors are associated with development of leukocytoclastic vasculitis. The table below includes the most common associations.

Histology

The full picture of LCV includes kariorrhexis, perivascular neutrophilic infiltrate, fibrin deposition, extravasation of red blood cells, endothelial swelling and dermal necrosis.

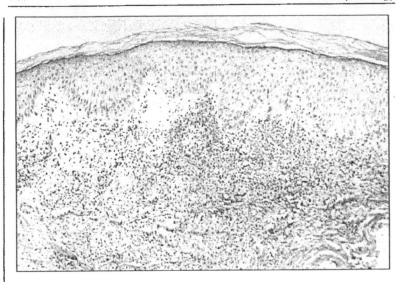

Fig 12.8. **Granuloma faciale**: Mixed dermal infiltrate with a grenz zone in the papillary dermis, abundant eosinophils, and leukocytoclasia.

12

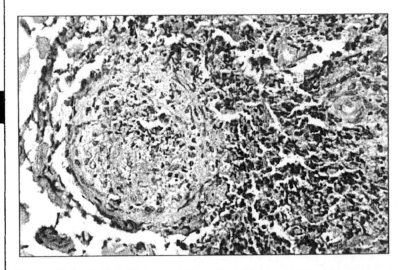

Fig 12.9. **Polyarteritis nodosa**: A mid-size artery shows necrosis of its wall with neutrophilic infiltrate, leukocytoclasia, and perivascular inflammation .

Henoch-Schonlein (Anaphylactoid) Purpura

Clinical

Clinically it is characterized by purpura, arthritis, abdominal pain and glomeru-lonephritis. It is the most common form of vasculitis in children. Although the etiology remains unknown, 50% of patients have a history of recent upper respiratory infection; organisms implicated are mycoplasma pneumoniae, Epstein-Barr virus, hepatitis B and others. Drugs and immunizations have also been implicated. Serum IgA and IgE are elevated in over 1/3 of the patients.

Histology

Histologically leukocytoclastic vasculitis is seen in the skin as well as in other organs. The changes can be mild and characterized by extravasated red blood cells. Immunofluorescence studies show IgA deposition in vessel walls of early lesions. This finding is diagnostic.

Urticarial Vasculitis

Clinical

It is seen in patients with chronic urticaria in which the clinical appearance is that of urticaria and/or angioedema with some ecchymosis.

Histology

Histologically the changes are those of urticaria in which usually discrete changes of vasculitis, particularly leukocytoclasia and nuclear dust are seen surrounding blood vessels of the dermis.

Pustular Vasculitis

Clinical

It is manifested clinically by papules, pustules, and areas of hemorrhage in addition to the purpuric lesions. Although pustular vasculitis is often a nonspecific manifestation of vasculitis, the term has been associated with specific conditions including Behcet's syndrome, bowel bypass syndrome, disseminated gonococcemia, and others.

Erythema Elevatum Diutinum

Clinical

Clinically it is characterized by persistent, red to violaceous papules, plaques, and nodules particularly on extensor surfaces of wrist, ankles, and other locations.

Histology

Histologically, the diagnostic findings are those of a leukocytoclastic vasculitis seen in earlier lesions, with a perivascular neutrophilic infiltrate, leukocytoclasia and fibrin deposition ("toxic kyalin") around blood vessel walls. In addition to neutrophils, there are histiocytes, lymphocytes and eosinophils. Older lesions may show

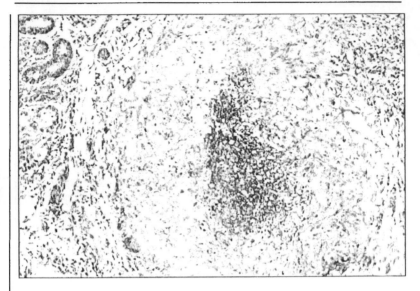

Fig 12.10. **Wegener's granulomatosis**: A granuloma with basophilic necrosis and mixed inflammation is seen in the dermis.

fibrosis of the dermis and macrophages with foamy cytoplasms so called extracellular cholesterolosis (Fig. 12.7).

Granuloma Faciale

Clinical
Clinically it is characterized by brown or red plaques or nodules on the face, persistent, not related to known causes.

Histology
Histologically there is a marked dermal infiltrate that is characteristically separated from the epidermis by a narrow grenz zone of uninvolved dermis. The infiltrate is polymorphous with an abundance of eosinophils together with neutrophils, lymphocytes, histiocytes and some plasma cells. Some degree of vasculitis, particularly deposition of fibrin material in and around vessel walls is seen. Extravasated red blood cells, and hemosiderin deposition are also seen (Fig. 12.8).

Large Vessel Vasculitis

Polyarteritis Nodosa

Clinical
It is an inflammatory process of small and medium sized arteries with involvement of many organs such as kidneys, nervous system, liver, and gastrointestinal

tract as well as the skin. Lungs are not involved. The prognosis is variable and depends on the visceral involvement. In relation to the skin, polyarteritis nodosa has been classified in two forms: a limited cutaneous form and a systemic form. The cutaneous form of polyarteritis nodosa is characterized by involvement of one or more medium sized arteries usually in the subcutaneous tissue of the legs with associated hemorrhage and occasional ulceration. In contrast the skin manifestations of the systemic form of polyarteritis nodosa are those of leukocytoclastic vasculitis, nondistinguishable morphologically from other causes of necrotizing vasculitis.

Histology
Histopathologically the cutaneous form of polyarteritis nodosa can not be easily differentiated from so-called nodular vasculitis, a morphologic subtype of erythema induratum, in which medium sized arteries of subcutaneous tissue or lower dermis are involved. In the latter, however, the changes of panniculitis are usually prominent although well formed caseating granulomas, a feature of classic erythema induratum, are often absent.

Kawasaki Syndrome

Clinical
Kawasaki syndrome (mucocutaneous lymph node syndrome), affects mainly children and infants, an acute multisystem disease characterized by:
1. fever unresponsive to antibiotics
2. bilateral conjunctival injection
3. oral changes including red lips, strawberry tongue or red oral pharyngeal mucosa
4. red palms and soles with edema of the hands and feet and desquamation of fingers and toes
5. rash
6. cervical lymphadenopathy.

12

The most serious manifestation is vasculitis of large coronary arteries. Untreated it may result in formation of coronary aneurysms with a mortality of 1-2%. Other arteries of spleen, kidneys, lungs and other organs may also be involved.

Granulomatous Vasculitis

Clinical
Allergic vasculitis (Churg-Strauss syndrome, allergic granulomatosis) is characterized by pulmonary and systemic vasculitis, extravascular granulomas and eosinophilia, occurring almost exclusively in patients with a history of asthma. The most common skin findings are palpable purpura, urticarial papules, nodules and ulcers as well as livedo reticulares.

Histology

Histologically, the involved areas of skin may show a leukocytoclastic vasculitis in which abundant eosinophils are seen. Other findings include the so-called pallisading neutrophilic granulomas in which macrophages are surrounding an area of basophilic necrosis, among which neutrophils with nuclear dust and eosinophils are seen. These extravascular granulomas, which were initially referred to as Churg-Strauss granulomas have been associated with other conditions including systemic lupus erythematosus, rheumatoid arthritis, Wegener's granulomatosis, polyarteritis nodosa, Takayasu's arteritis among others (Fig. 12.10).

Wegener's Granulomatosis

Clinical

Wegener's granulomatosis is characterized by necrotizing granulomatous vasculitis affecting the upper and lower respiratory tract, glomerulonephritis, and variable degrees of small vessel vasculitis involving multiple organs including the skin. On the skin, the lesions are variable including purpura and papulonecrotic lesions, particularly on extremities and buttocks. Serum antibodies of IgG class that react against cytoplasmic components of neutrophils are found in this entity. Two types of antineutrophilic cytoplasmic antibodies (ANCA) are found. A cytoplasmic "c" and a perinuclear "p". Most patients with Wegener's granulomatosis show positive antibodies, particularly the "c" type.

Histology

Histologically the lesions in the skin can appear as a necrotizing leukocytoclastic vasculitis, as a granulomatous vasculitis with angiocentric granulomas, and as extravascular pallisading neutrophilic granulomas (see above).

Temporal Arteritis (Giant Cell Temporal Arteritis)

It is a granulomatous vasculitis with a preference for the involvement of the superficial, temporal and ophthalmic arteries. Patients have severe headache, jaw claudication, and visual and neural disturbances.

Histology

The arterial wall shows a granulomatous arteritis involving the lamina media. Giant cells phagocyte the internal elastic lamina, which appears disrupted or even absent in some areas. Other inflammatory cells including lymphocytes, macrophages, and even eosinophils are seen.

Lymphomatoid Granulomatosis

Although often characterized as granulomatous vasculitis involving the lungs and other extrapulmonary locations, it is probably an angiocentric T-cell lymphoma.

CHAPTER 13

Dermal Deposits, Collagen Abnormalities and Calcifications

Ramón L. Sánchez and Sharon S. Raimer

Section 32. Dermal Edema

Edema of the dermis implies accumulation of serum or water between the collagen bundles. Although some degree of edema accompanies every inflammatory process, in some conditions the edema is quite prominent and is one of the key diagnostic features. The distribution of the edema can be superficial, involving predominantly the papillary dermis or it can be deep involving papillary and reticular dermis. When the edema of the papillary dermis is very prominent, the end result is a subepidermal blister. It is important to distinguish histologically the subepidermal blisters formed by separation of the basal membrane such as in bullous pemphigoid, from those entities in which the blister is due to edema of the papillary dermis.

Superficial Dermal Edema

Fixed Drug Eruption

Clinical
It is characterized by one or more, usually large, bullae, which develop after taking certain medications such as sulfonamides, tetracyclines, phenolphthalein and analgesics. Old resolved fixed drug areas show hyperpigmented macules.

Histology
Early lesions show a lichenoid reaction with hydropic degeneration of the basal layer of the epidermis and marked edema of the papillary dermis, sometimes with bulla formation. The infiltrate is mainly lymphocytic, but eosinophils are usually found. Neutrophils are sometimes present. The infiltrate extends up to the basal layer of the epidermis and occasional necrotic keratinocytes may be seen. Old, resolved lesions show mainly pigment incontinence with little inflammation and no dermal edema (see also Section 11, Chapter 6).

Differential Diagnosis
Erythema multiforme shows more necrotic keratinocytes, less eosinophils and the infiltrate is less lichenoid and more perivascular. Erythema dyschronicum perstans and resolved lichen planus show similar changes to resolved fixed drugs.

Table 13.1. Dermal edema

Superficial
Fixed drug eruption
Erythema multiforme
Lichen sclerosus et atrophicus
Polymorphous light eruption
Contact/Photo contact dermatitis
Photo drug reaction
Insect bite
Burns
Full Dermis
Urticaria
Insect bite
Ecthyma

Full Dermis Edema

Urticaria

Clinical

Urticaria is characterized by erythematous, edematous whelps on the extremities, trunk, and face, which are very pruritic and move in periods of less than 24 hours. They may appear target-like and are difficult to differentiate form erythema multiforme. Palmar and plantar lesions are common. Lesions of urticarial vasculitis are persistent, and may be purpuric.

Histology

The changes of urticaria are subtle and characterized by edema throughout the dermis manifested by separation of the collagen bundles, and by a certain pallor. There is a mild infiltrate throughout the dermis which is perivascular and interstitial. The infiltrate is always lymphocytic but, in addition, neutrophils and eosinophils can also be seen (Fig. 13.1). When the number of eosinophils is abundant, urticarial drug eruption is suspected. Urticarial vasculitis shows, in addition nuclear dust and extravasation of red blood cells around the superficial capillary plexus; however, the changes of vasculitis can be quite subtle.

Differential Diagnosis

Insect bites, in the acute phase, also show full thickness dermal edema and a mixed infiltrate with abundant eosinophils. Usually the infiltrate is more abundant than in urticaria and subepidermal blister formation may be present.

Section 33. Mucin Deposition

A small degree of mucin deposition is always seen in the dermis when special stains, such as colloidal iron or alcian blue, are obtained. The mucin is located generally near the epidermal and adnexal membranes. In some conditions, however, the deposition of mucin is so prominent that it is seen on H&E stains and constitutes the main histologic finding. It is helpful to classify the cutaneous mucinosis as focal

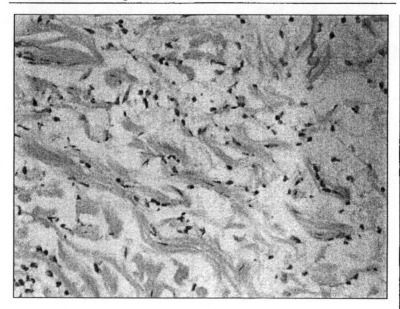

Fig 13.1. **Urticaria**: There is edema of the dermis and an interstitial infiltrate of lymphocytes, neutrophils and occasional eosinophils.

or diffuse. Cutaneous myxomas are probably benign neoplasias, but most mucinosis are passive accumulations of acid mesenchymal mucins, usually associated with inflammatory processes.

Focal Accumulations of Mucin

Mucous Cyst of the Oral Mucosa

Clinical

It is characterized by a small cystic lesion, usually located on the oral mucosa of the lower lip. If the cyst is intact, it should be translucent, but in a ruptured or inflamed cyst the appearance can be opaque and surrounded by erythema.

Histology

An intact cyst shows a well-delineated collection of mucin lined by macrophages, lymphocytes and plasma cells (Fig. 13.2). A minor salivary gland is usually seen in the vicinity and occasionally a communication between the gland and the cyst can be seen. In ruptured and inflamed lesions the mucin material is admixed with macrophages and acute and chronic inflammatory cells in variable amounts. In very inflamed cysts the mucin material may not be apparent.

13

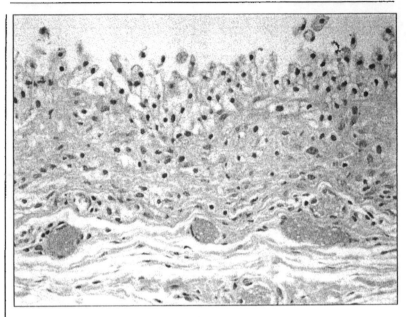

Fig 13.2. **Mucous cyst of the oral mucosa**: The wall of the cyst is made up of a collection of macrophages and foam cells with occasional lymphocytes.

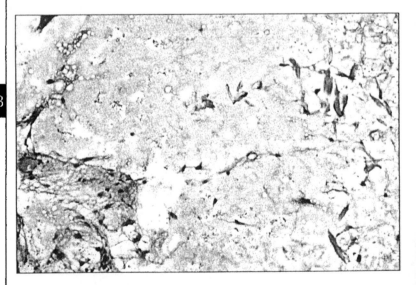

Fig 13.3. **Focal mucinosis**: The dermis shows an accumulation of basophilic, homogeneous mucin which separates the collagen bundles. Alcian blue stain.

Digital Mucous Cysts

Clinical

Digital mucous cysts appear clinically as tense, small cysts on the dorsal aspect of digits, often near the proximal nail fold.

Histology

Histologically, these show similar findings to the oral counterparts, except that the wall of the cyst may show only laminated collagen with few, if any, macrophages or inflammatory cells. The differentiation of digital mucous cysts from ganglion cysts is based on location and size.

Focal Mucinosis

Clinical

Focal mucinosis presents clinically as nondescriptive papules, usually on the extremities.

Histology

Histologically it is characterized by an accumulation of mucin in the mid and upper dermis. There is usually not a prominent cellular response, and the mucin often dissects the collagen bundles, presenting as small pools (Fig. 13.3). Focal mucinosis should be differentiated from cutaneous myxomas or angiomyxomas, tumors characterized by well-delineated dermal proliferation of stellate dendritic cells admixed within a mucinous stroma. Multiple capillaries are usually part of these tumors, contrary to myxomas in other organs, which are usually avascular. Reticulin stain shows a large number of reticulin fibers in these tumors, and the dendritic cells stain positive with factor XIIIa. Although angiomyxomas of the skin are often solitary, patients with multiple angiomyxomas are occasionally seen. Those patients may have Carney's syndrome, which includes lentigines, endocrine disturbances with adrenal over-activity and myxomas of the heart, in addition to cutaneous myxomas.

Alopecia Mucinosa

Clinical

It usually presents as an erythematous, somewhat elevated plaque with a smooth surface. Patients with extensive involvement, and those patients with associated mycosis fungoides have multiple small papules distributed throughout the trunk and extremities. Alopecia mucinosa is often seen in young patients without any systemic connotations. Older patients may present with alopecia mucinosa as part of cutaneous T-cell lymphoma, i.e., mycosis fungoides. Sometimes follicular mucinosis is the first manifestation of cutaneous lymphoma.

Histology

The hair follicles show accumulations of mucin that can be focal or can displace the cells of the follicle and also involve the sebaceous gland. Usually the mucin does not

13

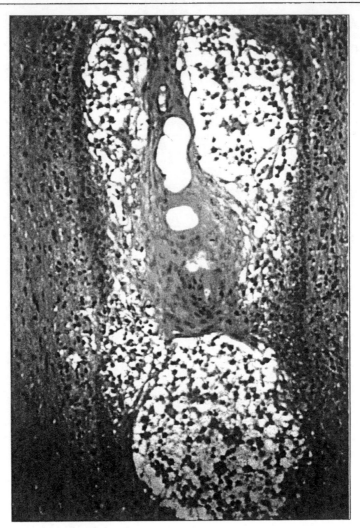

Fig 13.4. **Alopecia mucinosa**: The follicular epithelium is separated by clear spaces containing mucin. There is exocytosis of the lymphocytes.

spill into the surrounding dermis. Alcian blue and colloidal iron stains are usually used to confirm the presence of mucin. Mucicarmin stain also can be positive suggesting an epithelial origin. In cases of follicular mucinosis associated with mycosis fungoides, there is a marked infiltration of lymphocytes that have at least some degree of atypia, and that penetrate the hair follicles (epidermotropism) with formation of microabscesses. The epidermis may show also epidermotropism, but occasionally mycoses fungoides is almost exclusively folliculotropic.

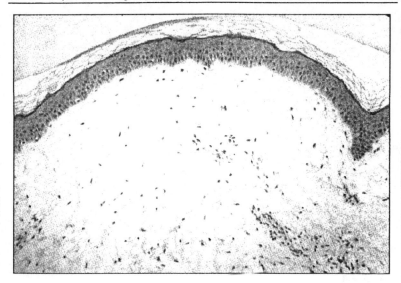

Fig 13.5. **Myxedema**: The dermis contains a large amount of mucin, resulting in a clear appearance. The number of cells is not increased.

Multiple/Extensive Mucin Depositions

Myxedema, Generalized and Pretibial

Clinical
Generalized myxedema is usually not clinically apparent, except perhaps as manifested by pale dry skin. Pretibial myxedema is usually characterized by unilateral or bilateral plaques on pretibial areas. Both conditions can be associated with thyroid disturbances, both as part of hyperthyroidism and with hypothyroidism.

Histology
Generalized myxedema shows a discrete increase of mucin throughout the dermis with only a small increase in dermal cells. Pretibial myxedema shows a prominent increase in mucin particularly in the deeper half of the dermis, as well as increased cellularity particularly by spindle cells, most likely fibroblasts (Fig. 13.5).

Lichen Myxedematosus (Papular Mucinosis)

Clinical
It shows multiple, 2-4 mm waxy papules, widely distributed, although characteristically seen on the dorsum of the hands, extremities and trunk. Serum immunoelectrophoresis shows a spike of monoclonal immunoglobulins, usually gamma. Patients with papular mucinosis may have also hoarseness, upper GI symptoms with difficulty in swallowing, cardiovascular abnormalities and CNS ischemic symptoms.

13

Histology
Histologically papular mucinosis shows an increase of mucin and cells in the upper dermis, quite well delimited. Colloidal iron and alcian blue stains help confirm the diagnosis.

Reticular Erythematous Mucinosis

Clinical
Reticular erythematous mucinosis usually shows a reddish patch or flat plaque on the sternal area.

Histology
Histologically there is a mild increase of mucin throughout the dermis, as confirmed with special stains.

Full Dermal Mucinosis

Scleredema of Buschke and Lowenstein

Clinical
Three main settings:
1. associated with diabetes mellitus;
2. following infections, particularly strep throat;
3. idiopathic.

The usual presentation is that of hard, indurated and thickened skin of the shoulders and upper back. It may resolve after a few years.

Histology
The dermis is thickened, as manifested by the sweat glands being quite high and apart from the subcutaneous tissue in what appears to be mid-dermis. The collagen bundles are separated by spaces that presumably contained mucin, which may have dissolved during processing. Frozen sections have been used in an attempt to preserve the mucin.

Lupus Erythematosus
Lupus erythematosus characteristically shows an increased amount of mucin in the dermis. The amount of mucin is diffusely distributed, can be seen on H&E stains and its presence is very significant in diagnosing this condition.

Section 34. Colloid, Amyloid and Other Deposits

Colloid and amyloid are seen on H&E as deposits of a homogenous eosinophilic or pale material, which is structureless, and is distributed in variable amounts in the papillary dermis or throughout the dermis. Both colloid and amyloid share some histochemical properties and they may be difficult to differentiate from each other. With electronmicroscopy, amyloid is formed by incomplete immunoglobulin chains and it is seen as microfilaments. Colloid, on the other hand, is seen as larger, less

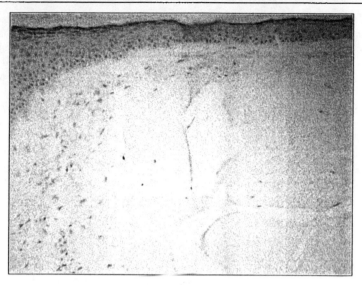

Fig 13.6. **Colloid millium**: There are large nodular collections of homogeneous eosinophilic/basophilic material in the papillary dermis with cleft formation.

homogenous filaments. Histologically, colloid has to be differentiated from solar elastosis, which it can resemble.

Colloid Millium

Clinical

Is clinically seen as nodular deposits with a waxy appearance on the dorsal hands and forearms, face, and other sun-exposed areas, particularly in elderly patients. There is also a juvenile presentation of colloid millium which is not related to long-term sun exposure. In both instances the clinical lesions are persistent.

Histology

Colloid millium exhibits deposits of a homogenous, pink, pale material in the papillary dermis. Because of the tissue processing, it is common to see artifactual clefts in the colloid nodules. The homogenous material is positive with PAS and also with amyloid stains such as congo red and crystal violet (Fig. 13.6).

Differentiation from nodular amyloidosis is usually by clinicopathologic correlation. A definitive distinction can be made by electronmicroscopy.

Nodular Amyloidosis

Clinical

Nodular amyloidosis is characterized by dermal nodules on variable locations. It represents skin deposits as a part of systemic amyloidosis or amyloid light chain (AL) protein.

Histology

It shows large deposits of a homogeneous eosinophilic material throughout the dermis without any specific architectural arrangement. Plasma cells may be located at the periphery of the deposits. The diagnosis is made with special stains, particularly congo red, which shows apple green birefrigence of the positive areas under polarization.

Lichen amyloidosus in contrast, shows only very small deposits of amyloid in the papillary dermis just below the basal membrane. The deposits may be so small that they may not be apparent histologically, even with special stains. In that situation, the diagnosis of lichen amyloidosus is suspected clinically because of the characteristic uniform slightly hyperpigmented appearing papules found most commonly on the lower legs. The amyloid molecule in lichenoid amyloidosis originates from the transformation of tonofilaments of keratinocytes in contrast to the amyloid in systemic or secondary amyloidoses (Fig. 13.7).

Lipoid Proteinosis

Clinical

It is a hereditary, autosomal dominant condition that is characterized clinically by waxy papules situated along the border of the eyelids (string of pearls), as well as by verrucous plaques on the elbows or knees. Mucosal involvement is manifested by hoarseness that can increase to difficulty in breathing. X-rays of the scalp show characteristic calcifications in the area of the hippocampus in the brain.

Histology

There is a deposit of eosinophilic homogeneous material, in the affected areas which surrounds in an onion skin fashion the basement membranes of the epidermis, adnexae, and blood vessels. The deposits may extend into the adjacent dermis (Fig. 13.8). The material stains positive with PAS after diastase ingestion and is made in part by reduplication of basal membrane material and by hyalin produced by fibroblasts.

Juvenile Hyaline Fibromatosis

Clinical

This rare congenital condition is characterized by the presence, since childhood, of multiple tumors located throughout the body that are hard in consistency, grow slowly and produce marked deformities.

Histology

The tumors are formed by spindle shaped cells surrounded by hyalin homogeneous material, which is positive on staining with PAS and alcian blue, although it is negative for amyloid stains. These hyalin deposits are found in the dermis, subcutaneous tissue, and occasionally in the submucosa of the oral cavity. Under electronmicroscopy the substance is formed by microfilaments.

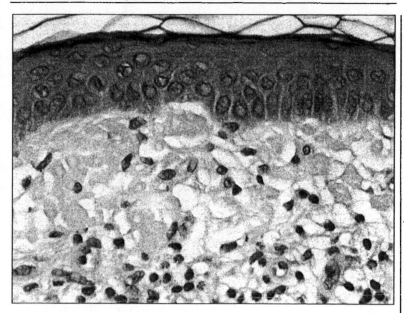

Fig 13.7. **Lichen amyloidosus**: The papillary dermis shows eosinophilic bodies admixed with a lymphoid infiltrate. Melanophages are also present.

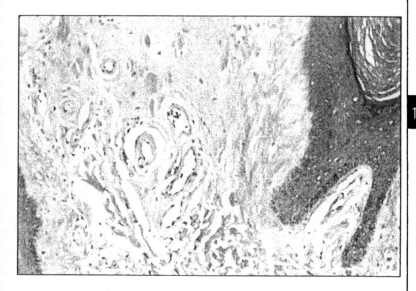

13

Fig 13.8. **Lipoid proteinosis**: The papillary dermis as well as blood vessels and adnexal basal membrane show deposition of eosinophilic, homogeneous material.

Table 13.2. Increased collagen, elastic fibers

Scar
Keloid
Morphea and scleroderma
Erythema ab igne
Elastosis perforans serpiginosa
Solar elastosis/Elastotic bands
Radiation dermatitis, late
Connective tissue nevus
Cutis vertices gyrata

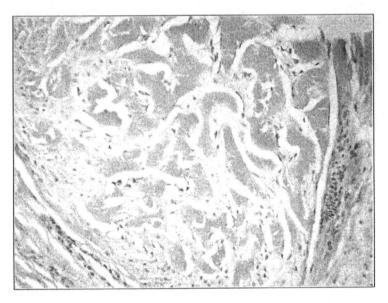

Fig 13.9. **Keloid**: Eosinophilic, acellular collagen bundles are seen among the fibrous stroma of a scar.

Section 35. Collagen, Elastic Fibers and Elastotic Material Abnormalities

Scar

Clinical

Scars result from damage to the dermis and/or subcutaneous tissue and are usually traumatic or iatrogenic. Many scars are small, often hypopigmented and appear slightly atrophic. Occasionally scars are prominent and protuberant and are usually termed hypertrophic scars. Keloids are seen in proned individuals, are more common but not exclusively in blacks, and are characterized clinically by extending to a larger size than the original skin injury.

Table 13.3. Decreased collagen, elastic fibers

Striae distensae
Mid dermal elastolysis
Aplasia cutis congenita
Anetoderma
Nevus lipomatosus
Focal dermal hypoplasia (Goltz syndrome)
Ehler's Danlos syndrome
Cutis laxa

Histology

Early scars are characterized by granulation tissue with acute and chronic inflammation, focal hemosiderin deposition and sometimes pigments resulting from cauterizing substances (such as Monsel's). Progressively, the granulation tissue is replaced by young collagen with decreased number of inflammatory cells. The end result is fibrosis with thicker collagenized bundles with decreased nuclei. The collagen bundles are criss-crossed but often run roughly parallel to the epidermis. In hypertrophic scars the amount of fibrous tissue increases considerably and is arranged in nodules of collagenized fibrous tissue. The histologic hallmark for the keloid is the presence of thick, acellular, hyaline-like bundles of collagen that run parallel and are surrounded by regular fibrous tissue (Fig. 13.9).

Morphea (Localized Scleroderma)

Clinical

The lesions of morphea are characterized by round or oval, indurated atrophic patches and plaques with a smooth surface and violaceous border. They can appear anywhere but are more common on the trunk and extremities. The head and neck are sometimes involved, although lesions of morphea are usually limited in extension. Some patients may present with multiple lesions and/or extensive, large, involved areas. Lesions of a linear type may be seen on the scalp (coup de savre) and the extremities. Superficial lesions which clinically resemble lichen sclerosus et atrophicus (LS&A) may be seen; in fact, LS&A has been defined by some as superficial morphea. Although both entities have several common characteristics, we prefer to keep them separated since LS&A is mainly a superficial dermatosis, while morphea characteristically involves deep dermis and subcutaneous tissue. Both entities are frequently seen as single processes independent from the other.

Histology

Early changes of morphea involve the subcutaneous tissue and deep dermis and are characterized by a lymphocytic infiltrate involving the fat and extending into the "lower" dermis. At this stage, therefore, the differential diagnosis should be established with paninculitis. The infiltrate characteristically extends into the eccrine glands which early on become atrophic. Plasma cells are sometimes an important component of the infiltrate. In well established lesions of morphea the

13

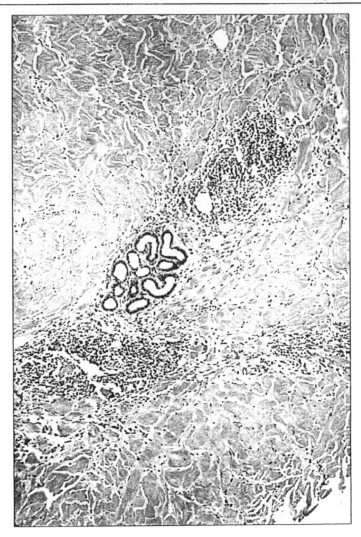

Fig 13.10. **Morphea**: A lymphoplasmocytic infiltrate in the subcutaneous tissue and lower dermis involves the eccrine glands. The dermis is thickened and fibrotic.

infiltrate has decreased, and there is fibrosis of the dermis with newly deposited collagen. The dermis appears hypocellular, the collagen bundles are homogeneous or hyalinized and thick fibrous septa extend from the dermis into the subcutaneous fat (Fig. 13.10). The new septa also appear sclerotic. In the sclerotic stage of morphea, low power examination of a punch biopsy shows characteristically straight, parallel, lateral borders with the deepest part of the biopsy being as wide as the epidermal surface.

System Sclerosis (Scleroderma)

Clinical

Scleroderma is a systemic disease, which, in addition to the skin, involves the gastrointestinal tract, lungs, kidneys and occasionally the heart. In the skin there are several manifestations including acrosclerosis, characterized by bound down skin of the hands and forearms, which become indurated and may lead to contractures and ankiloses of the joints. The face shows accentuation of the normal folds and decreased expression. Other skin findings include telangiectasias on the hands and nail folds, diffuse hyperpigmentation, and calcinosis cutis on the extremities. Raynaud's phenomenon is a common component of scleroderma. Esophageal involvement causes difficulty in swallowing.

A milder presentation of scleroderma with a better prognosis is CREST syndrome. It is characterized by C (calcinosis cutis), R (Raynaud's phenomenon), E (esophageal dysmotility), S (sclerodactyly) and T (telangiectasias).

Histology

The histologic findings in scleroderma are similar to those of morphea, particularly with fibrosis and sclerosis of the dermis. The diagnosis is made by correllating the clinical and pathological findings. Antinuclear antibodies are frequently present and the fluorescent pattern is often anticentromere.

Atrophoderma of Pasini and Pierini

Clinical

This is a subtype of morphea characterized by atrophic patches with a cliff-drop border.

Histology

Histologically the dermis may be of normal thickness or even of decreased thickness. Fibrosis and thick collagen bundles can also be seen.

Mid Dermal Elastolysis

Clinical

Clinically mid dermal elastolysis is characterized by wrinkling of the skin. It is usually seen in middle-aged women.

Histology

Histologically there is an absence of elastic tissue selectively in the mid dermis of involved areas. The elastic tissue around hair follicles is preserved.

Anetoderma (Macular Atrophy)

Clinical

Anetoderm is characterized by atrophic patches on the trunk which are felt as a hernial orifice when palpated. Some cases appear to be associated with a previous

13

inflammatory process (Jadassohn type) or lesions may be noninflammatory (Schweninger-Buzzi type).

Histology
Long standing lesions usually show a more or less complete absence of elastic fibers in the dermis. VVG stain or other stains for elastic fibers are needed to diagnose this condition.

Aplasia Cutis Congenita

Clinical
It is characterized clinically by a congenital absence of the skin at the time of birth. The process extends throughout the entire thickness of the dermis as seen histologically. Healing takes place after several months.

Alteration of Elastic Fibers

Pseudoxanthoma Elasticum

Clinical
Inherited condition transmitted as both an autosomal dominant or autosomal recessive disorder, characterized by abnormal elastic tissue seen on the skin, retina and blood vessel wall. The cutaneous manifestations appear after the second decade and are characterized by soft, yellowish papules on the neck, axilla and groin, giving a wrinkled or peau-d'orange appearance. Angioid streaks are seen in the eyes. Gastric hemorrhage may result from involvement of the vessels of the stomach, and coronary artery involvement may lead to angina pectoris. The arteries involved become calcified.

Histology
Examination of the skin on H&E shows largely increased, abnormal and bluish elastic fibers that appear fragmented and irregular. The bluish color of the elastic fibers is in part the result of calcification of the elastic fibers, which can be demonstrated with von Kossa stain.

Perforating Calcific Elastosis

Clinical
Presents clinically as periumbilical perforating lesions in middle-aged multiparous women.

Histology
Histologically, the findings are similar to those of pseudoxanthomaelasticum and are characterized by irregular, abnormal clumped elastic fibers that are also calcified and that are being extruded through the epidermis (transepidermal elimination). A histologic difference between these entities is the location of the elastic fibers, in mid

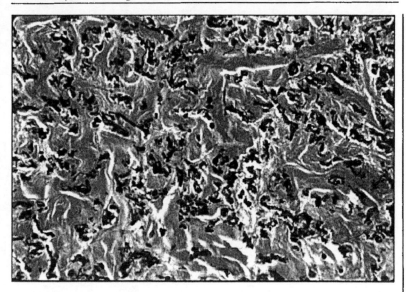

Fig 13.11. **Pseudoxanthoma elasticum**: The elastic fibers in the reticular dermis appear irregular, crumpled and fragmented. VVG stain.

Table 13.4. Calcium and cartilage depositions

Calcinosis cutis
Tumoral calcinosis
Idiopathic calcinosis of the scrotum
Osteoma cutis (Cutaneous ossification)
Malakoplakia
Chondroid syringoma
Accessory tragus

13

and deep dermis in pseudoxanthoelasticum, while they are superficial, including the papillary dermis, in perforating calcific elastosis.

Section 36. Calcification, Ossification, Cartilage Deposition

---

Calcinosis Cutis

Clinical

Calcinosis cutis refers to deposits of calcium in the skin and/or subcutaneous tissue. In general, calcium deposition follows hypercalcemic states (metastatic calcinosis) such as in hyperparathyroidism and/or hyperphosphatemia, usually secondary to renal failure. Other times, calcinosis cutis follows necrosis of the dermis (dystrophic calcinosis cutis) often associated with collagen vascular diseases such as

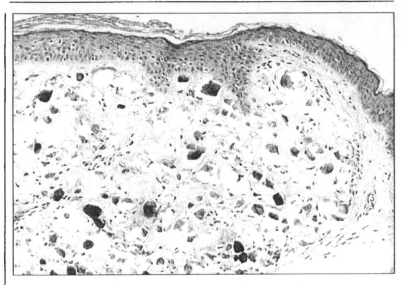

Fig 13.12. **Calcinosis cutis**: The upper dermis shows a well-delimited collection of calcium deposits among few inflammatory cells.

dermatomyositis, scleroderma and CREST syndrome, or lupus erythematosus. Idiopathic calcinosis cutis includes the familial condition known as tumoral calcinosis. There are also idiopathic calcinoses including idiopathic calcinosis resolution of the scrotum which may follow resolution of epidermoid cysts. When a single calcified nodule is found in the dermis without a known cause it is referred to as subepidermal calcified nodules.

Histology

Masses of calcium are seen replacing part of the dermis and/or subcutaneous tissue and appear a deep bluish to pink on H&E. Von Kossa stain will show the calcium deposits black. Surrounding the calcium deposits there is often a prominent foreign body type giant cell reaction which on occasion may be an indication of activity (Fig. 13.12).

Osteoma Cutis (Cutaneous Ossification)

Clinical

Cutaneous ossification can be seen as a primary event in Albright's hereditary osteodystrophy or may occur secondary to the ossification of a long standing inflammatory lesion or an area of calcinosis cutis. Several tumors show occasional ossification including basal cell carcinomas, intradermal nevi, chondroid syringomas, scars and nontumoral inflammatory conditions such as folliculitis, morphea, scleroderma and dermatomyositis.

Histology

Cutaneous ossification and osteoma cutis show spicules of bone in the dermis or subcutaneous tissue with osteocytes, cement lines, osteoblasts and osteoclasts.

13

CHAPTER 14

Panniculitis, Inclusions, Fungi and Parasites

Ramón L. Sánchez and Sharon S. Raimer

Section 37. Panniculitis

Panniculitis means inflammation of the subcutaneous tissue. The subcutaneous fat is roughly organized in lobules separated by thin fibrous septa. The blood vessels in the septa are usually larger than the small capillaries seen among the adipocytes. The fat is completely devoid of inflammatory cells, and any amount of inflammation in the subcutaneous tissue is abnormal. The hallmark of panniculitis is inflammation. Other findings are necrosis of the fat, which, on occasions, can be incomplete necrosis, without inflammation such as in the case of lupus panniculitis. Most of the time necrosis of the fat is complete and results in a marked inflammatory reaction with histiocytes, some with foamy cytoplasm, giant cells, lymphocytes, plasma cells and neutrophils. Typically panniculitis has been classified as being predominantly septal or lobular depending on which one of these compartments was primarily involved. The typical example of septal panniculitis is **erythema nodosum**, while for the lobular panniculitis the classic example is **erythema induratum**. In reality, if the process has been present for some time, both compartments are involved, and it is difficult to decide if an individual case is more septal than lobular. Another important histologic parameter is the presence or absence of vasculitis. Vasculitis, characterized by the presence of neutrophils in the wall of the blood vessels, with reactive endothelial hyperplasia and focal necrosis, may involve arteries, veins or venules.

Erythema Nodosum

Clinical

Bilateral, sometimes symmetrical, deep-seated nodules usually on the shins and/or thighs. They may be painful and appear reddish to purple in color. Erythema nodosum is a reactive process associated with multiple ethiologic factors. It often follows infections, particularly *beta hemolytic streptococcus* infections. It has been associated with tuberculosis, yersinia and deep fungal infections, such as blastomycosis and histoplasmosis. It has also been associated with some viral infections and with other conditions of unknown etiology, such as sarcoidosis and ulcerative colitis. Finally erythema nodosum is often associated with drugs, particularly sulfonamides, bromides and oral contraceptives.

Dermatopathology, edited by Ramón L. Sánchez and Sharon S. Raimer. ©2001 Landes Bioscience.

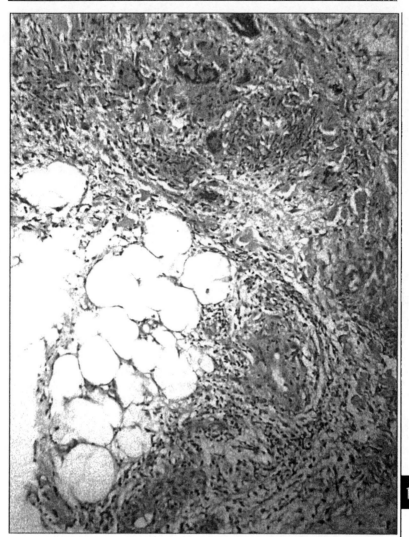

Fig 14.1. **Erythema nodosum**: Septal panniculitis with granulomata formation in the lower dermis

14

Histology

Erythema nodosum is a septal panniculitis, which implies that the septa are widened and show a variable infiltrate of neutrophils, lymphocytes, plasma cells and eosinophils (Fig. 14.1). From the septum the infiltrate will affect the lobules as well, so that a well-established erythema nodosum involves both compartments. Depending on the age of the infiltrate there will be a predominance of acute inflammatory cells or a predominance of lymphocytes and plasma cells. The lower dermis is generally also involved. Other important findings are granulomas, including Miescher's radial granulomas, usually not as numerous and with less necrosis than those seen with erythema induratum, and vascular involvement. The latter refers to involvement of veins without true vasculitis or involvement of arterial walls. Subacute migratory panniculitis is a form of erythema nodosum. In chronic erythema nodosum fibrosis and granulomas may be seen.

Erythema Induratum: Nodular Vasculitis

Clinical

Painful nodules predominantly in legs and thighs. They may be unilateral. Necrosis and drainage are possible.

Histology

Erythema induratum is a lobular panniculitis, from which it also involves the septae and lower part of the dermis. There is a mixed infiltrate with neutrophils, eosinophils, lymphocytes and plasma cells. Involvement of arterial walls, i.e., vasculitis, is often seen in erythema induratum and is a diagnostic feature that separates it from erythema nodosum (Fig. 14.2). Another important histologic feature is the presence of well-formed granulomas that often exhibit caseating necrosis.

Comment

Special stains for acid fast bacilli usually fail to disclose any micobacterial organisms in the granulomas of *E. induratum*. However, PCR studies have demonstrated micobacterial DNA among lesions of erythema induratum. Although this entity was classically associated with tuberculosis the connection between both entities was disregarded with the progressive disappearance of cases of tuberculosis. Nodular vasculitis refers to those cases of erythema induratum with prominent vasculitis and usually absence of granulomas. Histologically, cases of nodular vasculitis should be differentiated from panarteritis nodosa localized to the skin.

Other Panniculites

Pancreatic fat necrosis is associated with an acute or chronic pancreatitis. Histologically it is characterized by the presence of basophilic (chemical) necrosis of the fat, similar to the areas of fat necrosis seen around the pancreas in cases of pancreatitis.

Lupus profundus can cause areas of atrophy in patients with both cutaneous or systemic lupus. Histologically the fat is necrotic (absence of nuclei) and it shows an eosinophilic homogenous appearance (hyalinized). The collagen in the deep dermis may show basophilic necrosis, and focal vasculitis may be seen (Fig. 14.3).

Alpha-1 antitrypsin deficiency may present with draining sinuses in areas of panniculitis. The diagnosis is given by the presence of low serum titers of alpha-1 antitrypsin. A liver biopsy would show the characteristic eosinophilic cytoplasmic granules which stain positive with PAS.

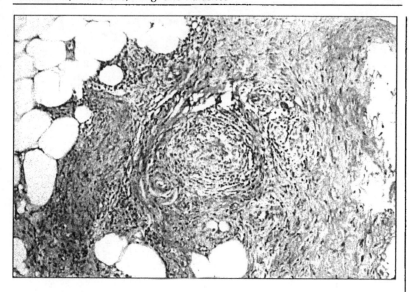

Fig 14.2. **Erythema induratum**: Predominantly lobular panniculitis with vasculitis involving mid-sized arteries.

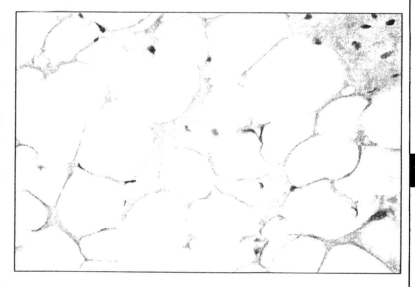

Fig 14.3. **Lupus profundus**: The fat appears acellular and eosinophilic. Other findings not seen include lymphoid follicle formation and neutrophils with leukocytoclasia.

Subcutaneous Fat Necrosis of the Newborn

Clinical

Characterized by focal areas of panniculitis, often on the face or upper trunk, sometimes associated with minor trauma. Histology: the fat shows crystals surrounded by multi-nucleated giant cells (Fig. 14.4). The crystals represent unsaturated fatty acids. Subcutaneous fat necrosis of the newborn should be distinguished from sclerema neonatorum, a severe condition with a very poor prognosis in which the fat appears bound down throughout the body.

Histology

Histologically in sclerema neonatorum multinucleated giant cells are replacing most of the fat.

Facticial panniculitis is usually caused by an injection of foreign material which triggers an acute and chronic inflammation of the fat. Examination of the involved subcutaneous tissue under polarization often shows birefringent crystals.

Cold panniculitis causes inflammation and necrosis of the fat associated with exposure to a cold environment. It has been described on the external thighs of women who have been horseback riding in the winter.

Section 38. Inclusions

Nuclear inclusions are characteristic findings in viral infections (see table). Cytoplasmic inclusions are seen in a variety of processes, including viral and bacterial infection, parasitized cells by protozoa or fungal organisms, and cytoplasmic components characteristic of specific entities.

Nuclear Inclusions

Herpes/Varicella Inclusions

Clinical
See page 44.

Histology
Formation of multinucleated giant cells is characteristic and diagnostic. The multiple nuclei appear empty with a ground glass appearance, with molding. In areas adjacent to the blister in the epidermis, some of the keratinocytes will show homogenization of the nuclei which acquires a bluish-gray color with margination of the chromatin. These characteristic viral inclusions are diagnostic for herpes.

Warts

Clinical
See page 19.

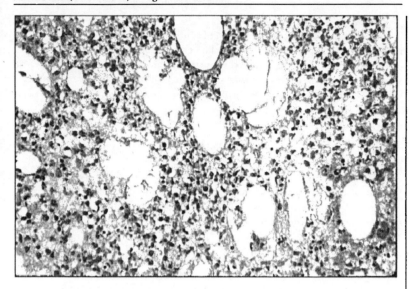

Fig 14.4. **Subcutaneous fat necrosis of the newborn**: Marked mixed infiltrate of the dermis. Remaining fat shows needle-shaped crystals with papillary arrangement.

Table 14.1. Nuclear inclusions

Molluscum contagiosum - Henderson/Patterson bodies
Herpes infection—Coudry type A inclusions
Cytomegaloviruses—Owls' eye
Orf/Milker's Nodules
Pseudo-inclusions in melanoma

Table 14.2. Intracytoplasmic inclusions

Molluscum contagiosum
Cytomegalovirus
Rhinoscleroma
Malakoplakia
Granuloma inguinale
Leishmaniasis
Histoplasmosis
Pneumocystis
Granular cell tumor
Recurrent infantile digital fibroma
Rhabdoid tumor of the skin

14

Histology

Inclusions in warts are occasionally seen as intranuclear basophilic bodies. There is hypergranulosis with thick keratohyaline granules either eosinophilic or basophilic, which should not be confused with cytoplasmic inclusions (myrmecia warts are characteristic), and also koilocytosis. The latter is characterized by a perinuclear halo in the keratinocytes of the upper part of the stratum of Malpighii, with irregular, angulated, relatively large nuclei.

Cytomegalovirus

Clinical

Cytomegalovirus is a very uncommon condition in the skin and is more frequently seen in other locations including the gastrointestinal tract and lungs. Occasionally, immunosuppressed patients show extensive areas of necrosis of the skin and subcutaneous tissue.

Histology

Histologic examination on those areas sometimes reveal typical cytomegalovirus inclusions usually in endothelial cells. Most of the time the inclusions are intranuclear, characterized by a large eosinophilic body surrounded by nuclear halo (owls eye). The necrosis of the skin and subcutaneous tissue may be the result of vascular damage by the virus. Occasionally cytomegalovirus shows also cytoplasmic inclusions manifested by eosinophilic, granular material.

Orf/Milker's Nodules

Clinical

These two entities are part of the pox DNA viral infections, the viral particles measuring 150-300 nm. Both entities are characterized by contact with affected animals (sheep or goats (Orf), cows (Milker's nodules)) or through fomites. Clinically, both conditions run a similar course characterized by the development of hemorrhagic macules that become papules, then nodules and involute spontaneously in three or four weeks.

Histology

The typical papule is characterized by hyperkeratosis and parakeratosis with deposits of eosinophilic material (viral inclusions) at the lower part of the stratum corneum. Some of the cells in the epidermis may appear swollen with clear cytoplasm. The dermis characteristically contains a marked vascular proliferation accompanied by prominent acute and chronic inflammation. As a result a misdiagnosis of granulation tissue or pyogenic granuloma/hemangioma is often rendered. The histologic diagnosis is suspected by the presence of the eosinophilic inclusion-like bodies in the stratum corneum, the clear appearance of the cells in the stratum of malpigium, plus the clinical history of contact with the animals. The diagnosis is confirmed by electromicroscopy which shows the oval shaped viruses in the stratum corneum and in the upper layers of the stratum of malpighium.

Table 14.3. Pox viruses

Cow Pox
Vaccinia
Variola (small pox)
Molluscum contagiosum
Milker's nodule (paravaccinia)
ORF (ecthyma contagiosum)

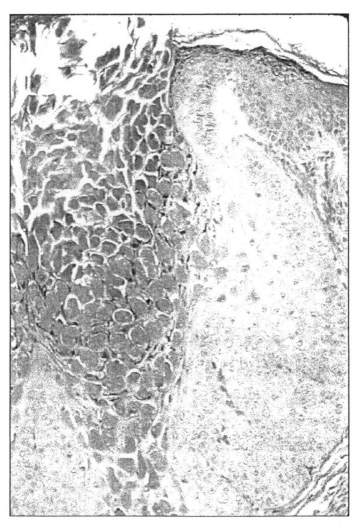

14

Fig 14.5. **Molluscum contagiosum**: The central core is filled with an accumulation of intracytoplasmic eosinophilic molluscum bodies (Henderson Patterson).

Cytoplasmic Inclusions

Molluscum Contagiosum

Clinical
It is characterized by the presence of 2-4mm papules with an umbilicated center often seen in children, although it can be seen at any age. Immunocompromised patients often show extensive lesions. Autoinnoculation is common.

Histology
Keratinocytes, particularly in the area of invagination, show large eosinophilic cytoplasmic inclusion bodies which displace the nucleus to the periphery, acquiring a signet ring appearance (Fig. 14.5). Old or irritated lesions of molluscum contagiosum may lack diagnostic inclusions and show only severe inflammatory reaction.

Leishmaniasis

Clinical
Leishmaniasis are rotozoal infections with worldwide distribution: **cutaneous leishmaniasis** (*L. Tropicana* and *L. Mexicana*), **mucocutaneous leishmaniasis** (*L. braziliensis*), **visceral leishmaniasis** and **kala-azar** (*L. donovani*). The skin lesions are characterized by papules, nodules and ulcers that may heal leaving a scar. A disseminated form with widespread nodules and macules without ulceration is also seen. *Espundia* refers to the destruction of cartilagenous areas in the nose, the floor of the mouth\ or tonsillar areas seen in mucocutaneous leishmaniasis due to **Leishmania braziliensis**. Leishmania parasites are transmitted by sand flies of the genus *Phlebotomus* and *Lutzomyia*.

Histology
The dermis is replaced by a severe infiltrate made up by lymphocytes, plasma cells, macrophages and epithelioid cells. Many of the macrophages contain parasites, which are seen as round to oval measuring 2-4 μm in size, and show a nucleus and an eccentrically located kinetoplast. Although they can be seen with H&E, they are better visualized with Giemsa stain. PCR techniques can be used in lesions suspected of Leishmaniasis with a low number of parasites.

Differential Diagnosis

Histoplasmosis
Organisms are about 3-5 μm in diameter, surrounded by clear capsule and absence of kinetoplast.

Rhinoscleroma
Mikulicz cells are large macrophages with clear cytoplasm that contain 2-3 um gram-negative rods, which are the causative organism *(Klebsiella rhinoscleromatis)*. The organisms can be seen with hemotoxylin and eosin, but are better visualized with Giemsa or silver stains.

14

Malakoplakia

Sheets of large histiocytes containing eosinophilic granules in their cytoplasm (von Hansemann cells). Also contain basophilic inclusions termed Michaelis-Gutmann bodies, 5-15μm in diameter. The Michaelis-Gutmann bodies, seen on H&E, are PAS positive and stain with von Kossa stain for calcium. They contain some iron stain as demonstrated with Perl's stain.

Granuloma Inguinale

Granuloma inguinale is caused by a small gram-negative bacillus (*Calymmatobacterium granulomatis*). On tissue smears stained by Wright or Giemsa methods there are large mononuclear cells containing red-stained intracytoplasmic encapsulated ovoid structures, so called Donovan bodies, 1-2μm in diameter.

Pneumocystis Carinii

Infection of the skin is very rare, although it has been described on AIDS patients. Silver stains show pneumocystis cysts among granular material in the dermis. Intracystic forms of the protozoa can be seen with Giemsa or Wright stains.

The cells of **a granular cell tumor** are round or oval. They display a regular nucleus and the cytoplasmic granules are PAS and S-100 positive. Ultrastructurally the granules are phagolysosomes. **Recurrent infantile digital fibroma**, which appears at birth or shortly thereafter, shows spindle shaped fibroblasts, some of which exhibit a paranuclear eosinophilic cytoplasmic body, 3-10 μm in diameter. Although the inclusion bodies are seen with H&E, they are more clearly demonstrated with phosphotungstic acid hemotoxylin and with Mason's trichrome. Ultrastructurally the tumor cells appear to be myofibroblasts and the inclusions bodies appear to be collections of myofilaments.

Section 39. Fungal and Fungal-Like Infections

Superficial infections by dermatophytes and yeasts are very common in dermatology and typically do not require a biopsy, since they are seen under microscope with potassium hydroxide preparations of skin scrappings (KOH). Atypical clinical presentations of tinea corporis and/or candidiasis, however, are sometimes biopsied. Deep fungal infections, on the other hand, are rare and can be seen in the skin as a primary infection, or the skin can be secondarily involved in systemic fungal infections. In this second instance the port of entry is often the respiratory system, and the patients are frequently immuno-suppressed.

Superficial Fungal Infections

Dermatophyte Infections

Clinical

The typical presentation is that of annular patches with a raised border and peripheral scale. In long-standing tinea there is frequent deep involvement of the dermis throughout the hair follicles, usually referred to as Majocchi's granuloma.

Fig 14.6. **Tinea corporis**: A PAS stain shows hyphae in the stratum corneum and a neutrophilic infiltrate is present throughout the epidermis.

Tinea of the scalp, as well as tinea of the beard, often involve the hair shafts, either penetrating it (endothrix) or the spores remaining between the hair shaft and the hair sheath (ectothrix). Tinea of the scalp often results in highly inflammatory lesions that can be destructive (kerion) and extensive (favus).

Histology

The histologic changes in tinea corporis may be minor and may require a high degree of suspicion. Sometimes hyphae can be seen on H&E in the stratum corneum. In other instances there are small subcorneal microabscesses with neutrophils. We routinely stain skin sections with PAS when we see subcorneal microabscesses. The upper dermis shows a mild infiltrate among which eosinophils are occasionally quite numerous. The diagnosis is given by the presence of hyphae seen with PAS or other fungal stains (Fig. 14.6). In the case of Majocchi's granuloma the hair shows an acute and chronic folliculitis with rupture. The surrounding dermis is severely inflamed with lymphocytes, occasional plasma cells, multinucleated giant cells, eosinophils and neutrophils. PAS stain often will show spores or hyphae, either in the hair follicle or outside in the dermis. Tinea capitis is usually well-demonstrated by examination of plug hairs with KOH which allows the recognition of spores around the hair follicle or in it. When a biopsy is examined the findings are similar to those in Majocchi's granuloma.

Candidiasis

Clinical

Often seen as part of intertrigo, candidiasis is also seen in bed-ridden patients on body surfaces that are occluded. In these situations the skin often shows small, moist, involved islands surrounded by normal skin.

Histology

Pseudohyphae and spores are seen in the stratum corneum, often associated with neutrophilic subcorneal microabscesses. As in the case of dermatophytes, PAS stains help to recognize the organisms. Candida infection is also seen frequently in necrotic skin, as in the case of burns or ischemic necrosis. In this situation candida organisms are seen almost exclusively in the necrotic tissues and not in the viable tissues. Invasive candidiasis is seen sometimes in immunosuppressed patients.

Pityrosporum colonization is seen clinically as a component of seborrheic dermatitis, and is also the cause of tinea versicolor. It is also seen in confluent and reticulated papillomatosis of Gougerot and Carteaud.

Histological: pityrosporum organisms are a common finding in the stratum corneum, particularly of the face, associated to hyperkeratosis, for instance in hyperkeratotic seborrheic keratosis or hypertrophic actinic keratosis. The organisms are seen on H&E as small yeasts. On PAS yeasts and hyphae are seen. Occasionally pityrosporum organisms are seen inside of hair follicles associated with acute inflammation, i.e., **pityrosporum folliculitis**.

Deep Fungal Infections

Clinical

They can be seen primarily in the skin or as secondary skin involvement in systemic infections. Skin findings include nodules or papules that are often ulcerated. The ulcers can be quite large, as in the case of sporotrichosis. When they involve an extremity lymphangitic spread is common (sporotrichoid spread). Several primary deep fungal infections on the skin present with verrucous appearance which can be clinically mistaken for keratoacanthoma or a wart. Examples of this type of presentation are **chromomycosis, blastomycosis**, and **coccidioidomycosis**.

14

Histology

Some of the deep fungal infections on the skin trigger a marked hyperplasia of the epidermis which appears extremely acanthotic and in which intraepidermal neutrophilic microabscesses are seen. The dermis underneath shows a mixed infiltrate with giant cells, macrophages, lymphocytes, plasma cells, neutrophils and eosinophils. Areas of necrosis are also seen. The characteristic organism that produces this histologic pattern is *Blastomyces dermatitis*, and the histologic pattern is called blastomycotic (Fig. 14.7).

Chromomycosis and **coccidioidomycosis** often show blastomycotic hyperplasia of the epidermis and epidermal microabscesses. Other nonfungal entities that present with this histologic pattern are blastomycosis-like pyoderma and **pemphigus vegetans**.

Characteristic of fungal infections of the skin is the so-called mixed granulomatous inflammation. In this, the infiltrate in the dermis presents not only multinucleated giant cells and histiocytes, but also a large number of lymphocytes, plasma cells, neutrophils and occasional eosinophils. Although clearly a granulomatous inflammation, truly well-formed granulomas are not seen except perhaps surrounding focal areas of necrosis. This barrage of cells is characteristic of fungal infection.

Chromomycosis

It belongs to the dematiaceous fungi, which produce a brown pigment that can be seen on H&E and stains positive with Fontana-masson, similar to melanin. Other related fungi in this category are the pheohyphomycosis (Fig. 14.8).

In tissue sections the organisms of chromomycosis are seen as brown yeast cells (copper pennies) that divide by a partition producing a central septum (medlar bodies). The organisms measure 6 to 8 microns and are seen loose among the inflammatory cells or inside giant cells.

Blastomycosis

The organisms are seen as round to oval spores measuring between 5 and 8 microns that occasionally show a polar body. They are better demonstrated with PAS, GMS or other fungal stains, and are seen as loose bodies or are found inside giant cells.

Cryptococcosis

Cryptococcosis produces two types of reactions depending on the grade of immunity in the host. In patients in which the immune system is relatively preserved, cryptococcosis presents as a granulomatous reaction, with well-formed granulomas. The organisms are seen in focal areas of necrosis or in the cytoplasm of giant cells. Immunosuppressed patients, on the other hand, present the gelatinous form of cryptococcosis characterized by multiple organisms in the dermis and subcutaneous tissue surrounded by abundant mucinous material. Inflammatory cells, particularly lymphocytes and neutrophils, are also seen (Fig. 14.9). The yeast cells as seen on H&E are small, about 5-7 microns round in diameter surrounded by a gelatinous capsule. The organisms stain black with fontana mason, while the capsule stains positive with acid mucin stains, such as alcian blue. In smears made from lesions of cryptococcosis, the organisms can be visualized by staining them with India ink, which affords rapid diagnosis.

Coccidioidomycosis

The inflammatory reaction in coccidiodomycosis is similar to that seen in blastomycosis and chromomycosis. The organisms are usually seen as large cysts, 20-30 microns in diameter, that are filled with small yeast cells, about 3 microns in diameter each (Fig. 14.10).

14

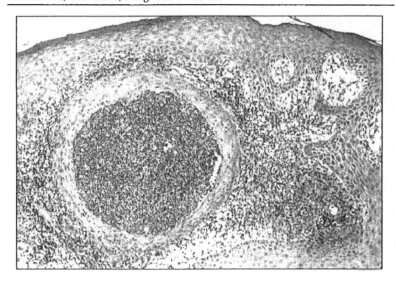

Fig 14.7. **Blastomycosis**: Epidermal pseudoepitheliomatous hyperplasia with formation of intraepidermal abscesses. Granulomatous dermal infiltrate with organisms.

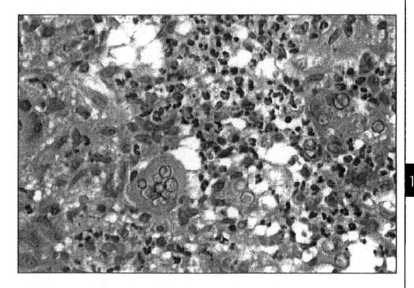

Fig 14.8. **Chromomycosis**: The dermis shows a mixed granulomatous inflammation with brown spores, some of them septated, in giant cells or free in the dermis.

14

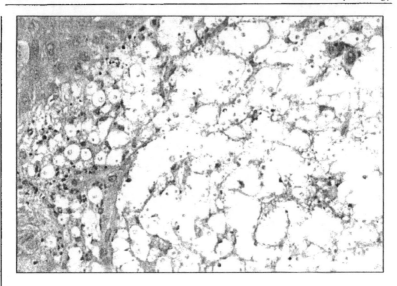

Fig 14.9. **Cryptococcosis**: Multiple encapsulated organisms are seen admixed in a mucinous stroma with little inflammatory reaction.

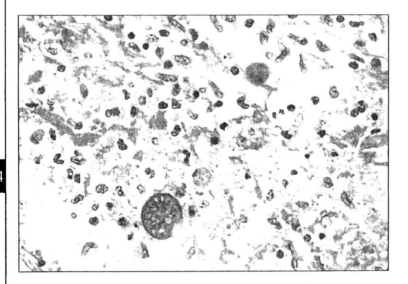

Fig 14.10. **Coccidioidomycosis**: A cyst with multiple endospores is seen in the lower part of the photograph. Lymphohistiocytic infiltrate is also present.

Sporotrichosis

Exhibits marked mixed granulomatous inflammation with extensive areas of necrosis. The organisms are seldom seen even after using special stains. When present they are small spores measuring about 6 microns in diameter. A characteristic finding is the asteroid bodies' formation which is diagnostic for sporotrichosis. Fungal cultures yield a high number of positive results. Because of the extensive necrosis and nonspecific chronic inflammation the differential diagnosis of sporotrichosis includes pyoderma gangrenosa.

Histoplasmosis

Histoplasmosis is very rarely a primary occurrence in the skin, where it is mostly seen as a manifestation of systemic involvement. The organisms are found mostly in the cytoplasm of macrophages that are infiltrating diffusely through the dermis. The *Histoplasma capsulatum* organisms are seen as small yeast forms measuring 4 to 6 microns in diameter, slightly oval and with a polar body seen clearly with PAS, which produces a "safety pin" appearance. In immunosuppressed patients histoplasmosis can present in the skin clinically as a follicular-like eruption which histologically resembles leukocytoclastic vasculitis, due to the presence of nuclear dust. In this latter instance the organisms are often seen extracellularly as well as in the cytoplasm of occasional macrophages present (Fig. 14.11).

South American Blastomycosis:

Paracoccidioidomycosis: Pilot's Wheel is the Characteristic Finding

Actinomycosis

Actinomycosis is not a true fungus but a filamentous bacteria. It is characterized by extensive areas of inflammation with areas of necrosis. The organisms are seen as bacterial colonies which appear clinically as yellowish grains. Histologically the bacterial colonies are seen as purple, round, structureless material with radial extensions at the periphery. The Splendora effect, an eosinophilic loose rim surrounding the colony, can be seen. Gram stain shows gram positive filamentous bacteria forming the bacterial colonies (Fig. 14.12).

14

Nocardiosis

Produced by a filamentous bacteria. Shows extensive inflammation, among which the organisms can be seen. The nocardia stains positive with fite stain, and therefore should be differentiated from mycobacterial infections. They also stain with gram stain as gram negative organisms. Actinomyces and nocardia organisms may produce the **actinomycetomas**. **Eumycetomas** are caused by true fungi including *Allescheria, Pseudoallercheria* and *Madurella* organisms.

Botryomycosis

In spite of the name it denotes a pyoderma with extensive granulation tissue that is usually seen on the scalp or extremities. It is generally caused by *Staphylococcus aureus*, and bacterial colonies can be seen histologically.

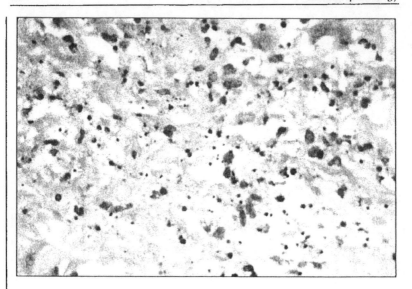

Fig 14.11. **Histoplasmosis** in immunosuppressed patient: Extracellular organisms are seen in conjunction with abundant nuclear debris.

14

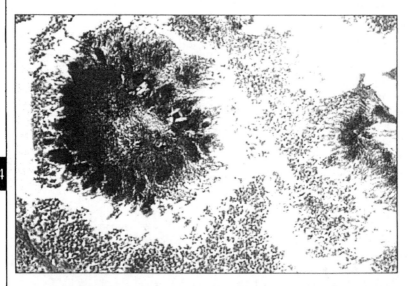

Fig 14.12. **Actinomycosis**: Colonies of elongated bacteria (sulfur granules) are surrounded by a prominent neutrophilic infiltrate.

Section 40. Cutaneous Parasitoses

Scabies

Clinical
Caused by the itch mite *Acarus scabiei*. It is characterized clinically by multiple, extremely pruritic papules on fingers, forearms, elbows, trunk, abdomen, external genitalia and thighs with the formation of burrows. The diagnosis is usually made by scraping off the mites and looking at the microscope with mineral oil preparation. Positive findings include mites, eggs and feces.

Histology
Occasionally biopsies are obtained of scabies lesions. The parasites or parts of it are usually seen in the stratum corneum. Immediately below there is an acute inflammatory reaction characterized by the presence of multiple eosinophils and some neutrophils and lymphocytes. Vascular fibrinoid necrosis may be seen focally. Farther away the dermis shows edema, lymphocytes, neutrophils and eosinophils. The Norwegian scabies patients have innumerable mites with formation of crusted papules in plaques. Histologically, a large number of mites are seen in the stratum corneum (Fig. 14.13).

Cutaneous Larva Migrans

Clinical
It is caused by the filariform larvae of the dog and cat hookworms. It is also called creeping eruption. Presents clinically with linear erythematous tracks where the parasite has been traveling.

Histology
Biopsy of those tracks are usually negative since the parasite moves continuously.

Strongyloidiasis

Clinical
It is caused by *Strongyloides stercoralis*, and is primarily a parasitosis of the small intestine. Re-infection can occur at the level of the anus with cutaneous manifestations on the buttocks, groins and lower abdomen caused by the rapid movement of the parasites on the skin (**larvae currens**). Systemic infections in immunosuppressed patients can be seen as a result of invasion of the blood stream by **filariform larvae**, which shows clinically a purpuric reticulate macular eruption, particularly on the abdomen.

Histology
Biopsies of the skin on the purpuric areas occasionally show larvae in dermal capillaries without inflammation. There is also extravasation of red blood cells.

14

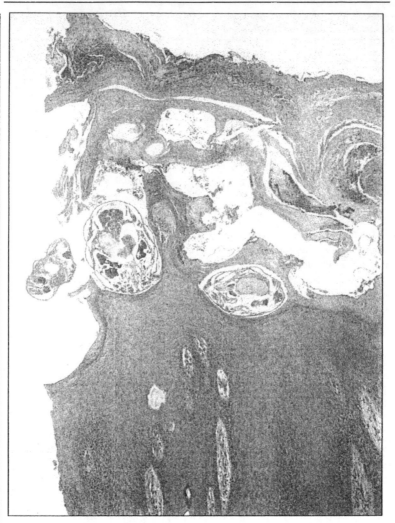

Fig 14.13. **Norwegian scabies**: The epidermis shows a thick stratum corneum in which multiple scabies mites are seen.

Onchocercosis

Clinical

Onchocercosis is transmitted by the larvae of *Onchocerca volvulus* introduced into the human body by black flies. On the skin the parasite matures to the adult stage in the subcutaneous tissue, forming symptomatic subcutaneous nodules called onchocercomas. The microfilaria produced by the adult worms live in the dermis and on the aqueous humor of the eyes.

Histology
Biopsy of the skin on these areas sometimes yields small larvae inside capillaries of the dermis without inflammatory reaction.

Subcutaneous Dirofilariasis

Clinical
Transmitted by mosquitoes, subcutaneous dirofilariasis is characterized by subcutaneous nodules.

Histology
Transverse and longitudinal sections of the worm, measuring between 125- 250 μm in diameter, can be seen. The usual host of the dirofilaria are dogs where it causes heartworm disease.

Cysticercosis

Clinical
The heart tapeworm *Taenia solium* develops in the human intestinal tract after ingestion of poorly cooked pork containing *T. solium* larvae. The eggs deposited by the larvae can be ingested by humans and the larvae born invade blood and lymph vessels and produce cysticerci in subcutaneous tissue, eyes, brain, skeletal muscle and the heart.

Histology
The cyst contains clear fluid and a membranous structure representing the larvae (**Cysticercus cellulosae**).

Schistosomiasis

Clinical
There are three species of schistosoma: *S. mansoni* in the Caribbean, *S. japonicum* in Eastern Asia and *S. hematobium* in the Middle East and Africa.

Histology
They are distinguished histologically by the location of the spine in the ova, in the apical position in *S. hematobium*, and lateral in *S. mansoni*. *S. Japonicum* does not have a spine.

14

CHAPTER 15

Hair Follicle Abnormalities

Ramón L. Sánchez and Sharon S. Raimer

Section 41. Folliculitis

Acute Bacterial Folliculitis

Clinical

Acute bacterial folliculitis are follicular-centered red papules with small pustules, generally superficial, often seen on the trunk, particularly the back, scalp, and proximal extremities associated with rubbing of clothes, scratching or shaving (legs of women). Shaving is also a common cause of folliculitis on the face in men. Another common area for folliculitis is the back of the neck and, on occasion, the inflammatory process extends deep into the follicle and even into adjacent follicles, forming a furuncle.

Histology

There is an acute, predominantly neutrophilic infiltrate filling the follicular opening and occasionally the acrotrichium and the intradermal hair follicle. Associated with the infiltrate there is necrosis and rupture of the follicular wall with eventual extension of the inflammatory process in the dermis (Fig. 15.1).

Note: When the histologic section does not include the hair follicle, one often sees an acute infiltrate in the mid dermis unrelated to other pathology. In those cases deeper sections often disclose an acute folliculitis.

Pityrosporum Folliculitis

Clinical

Predominantly in the back, it is characterized by a pruritic type of folliculitis that does not respond to antibiotic therapy.

Histology

Similar to bacterial folliculitis in which pityrosporum yeast forms are seen in the infundibulum of the hair follicle associated with acute inflammation. The organisms can be seen on H&E stains, but are easier to demonstrate with PAS stain. We perform PAS stains in all cases of acute folliculitis, regardless of the clinical diagnosis.

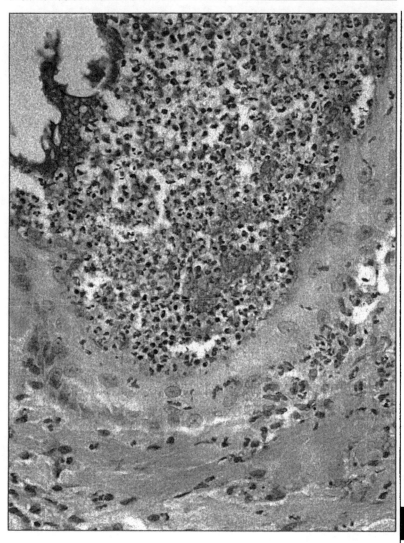

15

Fig 15.1. **Acute folliculitis:** Follicular opening is filled with neutrophils and necrotic debris. Neutrophils are seen also in the wall of the follicle.   .

Eosinophilic Pustular Folliculitis

Clinical

It is seen more commonly on the trunk and face, although it has been described on the extremities and even in glabrous skin. Often seen in AIDS patients, it can also be seen in immunocompetent individuals and in children.

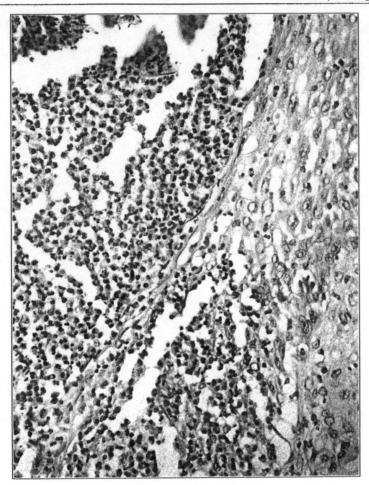

Fig 15.2. Eosinophilic pustular folliculitis: A follicular plugging is seen made up of
eosinophils which are breaking through the follicular wall involving the dermis.

15

Histology

There is an acute folliculitis, in which the infiltrate filling the hair follicles contains
a high number of eosinophils. The eosinophils extend also into the neighboring
dermis. No organisms are seen on special stains (Fig. 15.2).

Majocchi's Granuloma

Clinical

It occurs in deep-seated, long-standing tinea corporis. It is often seen on extremities, but it can be seen elsewhere. The skin shows an erythematous scaly patch

characteristic of tinea corporis in which eosinophilic small deep-seated papules are scattered throughout.

Histology

The dermis shows a marked mixed infiltrate which often can be seen centered around a ruptured hair follicle. The infiltrate is made of neutrophils, abundant eosinophils, multinucleated giant cells and occasional lymphocytes. The diagnosis is given by the presence of fungal spores or hyphae in the hair follicle and in the dermis. The diagnosis of Majocchi's granuloma can be suspected, but not definitively made, in the absence of fungal spores.

Other Folliculites

Acne cysts and acne folliculitis cannot be reliably differentiated from acute bacterial folliculitis except by history. **Furunculosis** shows a similar acute infiltrate that is more extensive and is associated with marked dermal necrosis. **Dissecting folliculitis** of the scalp shows, in addition to acute inflammation and necrosis, presence of a lymphoplasmocytic infiltrate and scarring of the dermis. **Acne necrotica** is a superficial folliculitis of the scalp.

Section 42. Hair Follicle Abnormalities

Alopecias, Nonscarring

Alopecia is the loss of hair, either as an acute episode or as a chronic hair loss. Alopecias can be localized, or they can be diffuse. In either one, the hair loss can be complete, or it can be partial, which is often referred to by the patient as thinning of the hair. Localized alopecias can also be patterned when they involve an area characteristic of certain types of hair loss, i.e., androgenetic alopecia. Another useful classification is that of nonscarring or scarring alopecias. From a clinical point-of-view nonscarring alopecias exhibit a smooth skin surface without any fibrosis or atrophy. Scarring alopecias are associated with fibrosis, whether extensive and involving a great part of the dermis, or discrete, manifested only around the involved hair follicles or replacing the atrophic follicles. Nonscarring alopecias eventually become scarring after the process has progressed enough to become irreversible.

Scalp skin biopsies are processed in two different ways, depending upon the clinical diagnosis as scarring or nonscarring alopecia. For scarring alopecias, the biopsies are processed in the traditional longitudinal way, in which the section runs perpendicular to the epidermis. In this way, only a few follicles are seen, but the inflammatory process and its relationship to the epidermis or to the hair follicle is easily seen. Longitudinal sections also offer a better view of the full dermis, and the presence of fibrosis. In nonscarring alopecias, it is probably better to use transverse sections or a combination of longitudinal and transverse sections. In transverse sections the cut surface runs parallel to the epidermis. An initial cut is performed at the level of the isthmus of the hair, at about one millimeter from the epidermis. In this way two cylinders result from the original punch biopsy. The top cylinder includes

15

epidermis and upper dermis, while the bottom one includes the rest of the dermis, as well as subcutaneous tissue. The histologic sections begin in both cylinders at the level of the isthmus of the hair follicle, the portion of hair situated between the sebaceous gland and the arrector pilaris muscle. The two adjacent surfaces where the cut section has been performed are inked and placed down, so that the serial and deeper sections of the two cylinders will go toward the surface in the upper cylinder and toward the subcutaneous tissue in the lower one.

The histologic examination of a transverse oriented biopsy of the scalp shows cross-sections of all the hair follicles present at that level. If it is at a superficial level, it will include not only terminal hairs, but also vellus hairs. In that sense it will be easy to count hair follicles and reach a conclusion if there is or there is not a decreased number compared with normal values. Hair follicles are grouped into follicular units, each one consisting of three or four terminal hairs, one or two vellus hairs, and the sebaceous gland. Altogether in a four millimeter punch biopsy there should be about six follicular units with a total of twenty to twenty-five terminal hairs and about six to eight vellus hairs. Most of these hairs will be anagen, with the bulb located in the subcutaneous tissue. At the isthmus, the inner root sheath disappears and the external root sheath starts to undergo epidermal keratinization. Catagen hairs are characterized by apoptosis and individual cell necrosis of cells in the root sheath, perhaps associated with a neutrophilic infiltrate. The inner root sheath disintegrates earlier in the catagen hairs. As a result, the hair shaft becomes soon in contact with the outer root sheath. Telogen hairs or club hairs are also seen in the upper levels, and they lack inner root sheath. The retraction scar on the telogen hairs is also commonly apparent. Vellus hairs undergo the same stages as terminal hairs (anagen, catagen and telogen) and are recognized by the diameter of the hair shafts, which is smaller than the thickness of the inner root sheath.

Alopecia Areata

Clinical
Alopecia areata is characterized by well-defined patches of alopecia with sharply delineated borders in any area of the scalp. Commonly seen in young patients, it can present with several patches or be extensive, involving the whole scalp (alopecia totalis), or even the scalp and other areas such as the eyebrows or beard (alopecia universalis). On close examination so called exclamation point hairs are seen.

Histology
The scalp transverse sections show a high number of catagen hairs and/or dysmorphic, abnormal hairs. A lymphocytic infiltrate may be seen around the lower portion of the hair follicle, and miniaturized hairs are sometimes seen (Fig. 15.3).

Androgenetic Alopecia

Clinical
Androgenetic alopecia shows loss of hair in frontal areas or on the crown. This type of alopecia is also seen in women, often as hair loss on top of the scalp.

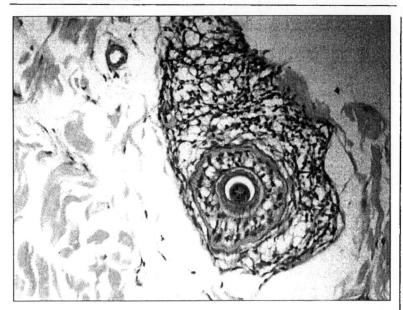

Fig 15.3. **Alopecia areata**: Miniaturized anagen hair is seen surrounded by a lymphocytic infiltrate (swarm of bees).

Histology

The transverse scalp biopsy shows a shift from terminal hairs to vellus hairs over the customary ten percent of vellus hairs normally present. Miniaturized hairs are also seen and are diagnostic for this type of alopecia (Fig. 15.4).

Trichotillomania (Traction Alopecia)

Clinical

It is characterized by well-defined patches of alopecia, often with irregular contours or incomplete balding.

Histology

Histologically the key finding is the presence of empty follicles with hemosiderin casts, evidence of traumatic hemorrhage. In later stages, or concurrently in chronic cases, there is a shift into telogen and early anagen hairs (Fig. 15.5).

Telogen/Anagen Effluvium

Clinical

Characterized by rapid shedding of a large number of hairs throughout the scalp, often associated with an illness, trauma, or traumatic experience. This is commonly seen in women after giving birth.

15

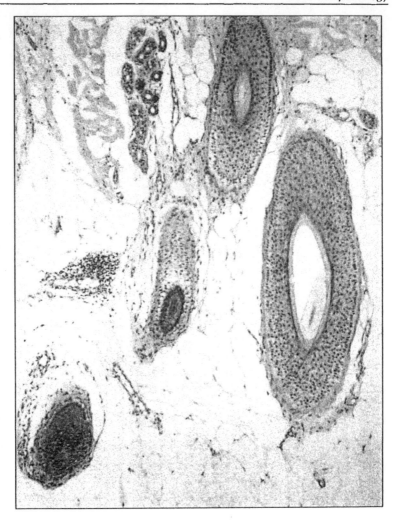

15 Fig 15.4. **Androgenetic alopecia**: Increased vellus hairs and miniaturized hair follicles without inflammation.

Histology

Histologically the transverse sections of the scalp biopsy show a considerable shift into the catagen/telogen stages. More than 12 percent of telogen hairs in a given biopsy is abnormal and highly suspicious of telogen effluvium.

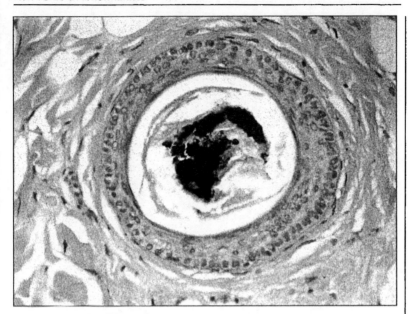

Fig 15.5. **Trichotillomania**: A melanin cast is seen inside an empty follicle.

Drug-Related Alopecia

Drug related alopecia, particularly alopecias associated with chemotherapy often show only empty follicles, so-called anagen effluvium. With time an involution of those follicles and formation of early anagen is the rule.

Alopecias, Scarring

To study scarring alopecias longitudinal sections of scalp biopsies are preferred.

Lichen Planopilaris

Clinical

Clinically characterized by inflammatory papules on the scalp associated with scarring.

Histology

Histologically there is a lichenoid infiltrate seen not only on the papillary dermis, but also surrounding the hair follicles. There is hydropic degeneration with separation of the follicular epithelium from the perifollicular dermis. Civatte bodies may be seen. The differential diagnosis includes discoid lupus erythematosus (Fig. 15.6).

15

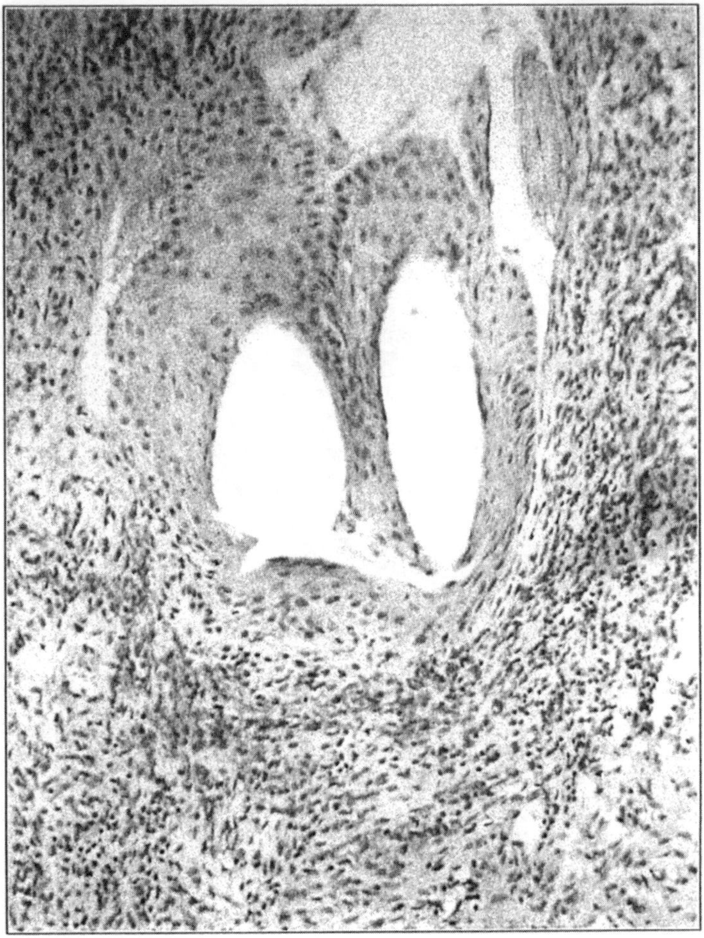

Fig 15.6. **Lichen planopilaris**: Scarring alopecia with a marked perifollicular, lichenoid infiltrate.

Pseudopelade of Brock

Clinical

It is clinically characterized by irregular patches of alopecia, with irregular contours resembling digital prints.

Histology

Histologically most biopsies are obtained in chronic long-standing cases, and the longitudinal sections of scalp biopsies are most often devoid of inflammatory changes and absence of hair follicles. Absence of sebaceous glands is also the rule, while the arrector pilaris muscles are still present in the dermis. In the areas where the hair follicles have involuted there is a scar replacing the previous follicle. These findings can be better outlined by an elastic tissue stain which shows absence of elastic fibers in the follicular scars, while the surrounding reticular dermis has a normal amount of elastic fibers. In pseudopelade there is no increase in elastic fibers in the dermis.

Folliculitis Decalvans

Scarring alopecia follows folliculitis decalvans. The findings are similar to those in necrotizing folliculitis of the scalp. Depending on the stage in which the biopsy was obtained, a marked infiltrate is seen replacing and involving the dermis. The infiltrate is both acute and chronic, and plasma cells are often numerous. Necrosis of hair follicles is the rule.

Occasionally patches of alopecia are secondary to metastatic carcinomas, particularly from the breasts and the lungs. **Alopecia syphilitica** is seen in secondary syphilis. Histologically there is an abundance of plasma cells. Organisms are rarely found.

Follicular Plugging/Hyperkeratosis (Comedones)

Clinically these are seen as dilated follicular openings that can be open or closed. In the latter case the follicular plug appears dark brown or black. Histologically the infundibulum of the hair follicle is greatly dilated and filled with keratin material admixed with sebaceous gland secretions.

Favre-Racouchot (Nodular Elastosis with Cysts and Comedones)

Clinical

It is seen mainly on the malar areas and near the external border of the eyes in older men. It is characterized by an irregular, papillary skin surface with multiple dilated cysts.

Histology

Multiple dilated comedones are seen, surrounded by dermis with prominent solar elastosis.

Milia Cyst

Clinically

These are clinically characterized by small, one to two millimeter, yellowish cysts seen near the surface of the epidermis. They occur most commonly on the face.

Histology

Histologically there is a small epidermoid cyst (Fig. 15.7).

Other Related Entities

Acne cysts are seen clinically as inflammatory papules. Histologically the hair follicles appear dilated and filled with keratinous debris, usually admixed with an

15

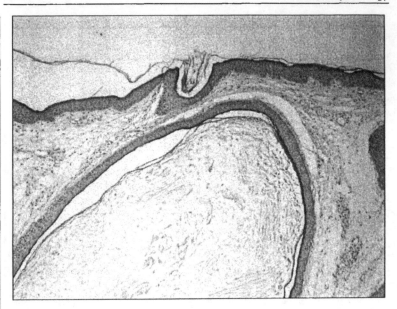

Fig 15.7. **Milia cyst**: Small epidermal cyst on the upper dermis filled with layered keratin.

inflammatory infiltrate. The surrounding dermis usually shows a prominent and acute inflammation which involves part of the follicular wall.

Trichostasis spinulosa is seen clinically as dilated hair follicles from which a tuft of multiple vellus hairs, 15-20, can be seen with the aid of magnification. Histologically the follicle is dilated, and if the cut section allows, a tuft of vellus hairs are seen in the follicular opening.

Keratosis pilaris is clinically characterized by keratotic follicular papules seen mainly on the shoulders, arms, and thighs. Keratosis pilaris can also be seen on the face and on other areas of the body. Histologically the follicular opening is dilated and shows prominent hyperkeratosis.

Nevus comedonicus is clinically characterized by a group of comedones that can be seen in an annular or serpiginous distribution. Often seen on the hands, they can be seen elsewhere, particularly on the head and neck. Histologically dilated comedones without inflammation are seen.

Perifollicular Inflammation

Acne Rosacea

Clinical

Acne rosacea is manifested most commonly by erythematous papules that are particularly seen in a malar distribution and around the nose. Telangiectasiae are commonly seen in association with the erythema. A manifestation of acne rosacea localized to the nose is sebaceous hyperplasia, which can be so prominent as to produce deformities of the nose with considerable increase in size. This last manifestation is called **rhinophyma**.

Histology

The most common histologic manifestation of acne rosacea is perifollicular inflammation, made up mainly by lymphocytes. Occasional plasma cells can be seen. The inflammation often involves the wall of the hair follicle. Compact granulomas with small areas of necrosis can be seen also in the upper dermis in proximity to the hair follicles (Fig 15.8). These granulomas cannot be distinguished on morphologic grounds from granulomas seen in mycobacterial infections and on sarcoidosis. Almost invariably special stains are negative. Telangiectasia and sebaceous hyperplasia are also seen in this entity, as described clinically.

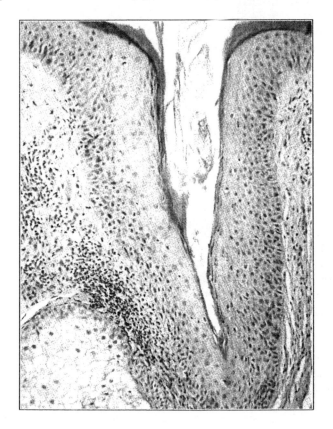

15

Fig 15.8. **Acne rosacea**: Perifollicular inflammation with demodex folliculorum mites in the follicular opening.

CHAPTER 16

Cysts, Polyps and Hamartomas

Ramón L. Sánchez and Sharon S. Raimer

Section 43. Cysts

A cyst is an epithelial-lined space in the skin. The lining of the cyst may be derived from one of many different sources, including eccrine ducts, hair follicles, and sebaceous ducts. The contents of the cysts are variable depending on the tissue of origin, often being made by keratin.

Epidermal Inclusion Cysts

Clinical

By far the most common type of cyst of the skin, epidermal inclusion cysts occur anywhere, although less commonly in glabrous skin. Usually asymptomatic, they may increase to a considerable size (four and more centimeters). Very small cysts, as seen often on the face, are denominated milium (or milia). Some cysts communicate with the surface, with periodical drainage of the cyst contents through its opening. When the wall of the cyst ruptures a marked inflammatory reaction ensues, manifested by redness and swelling of the skin overlaying the cyst, pain and liquefaction of the cyst contents. If left untreated a draining sinus is formed eventually, through which the inflamed and necrotic material is extruded.

Histology

Epidermal inclusion cysts are lined by a type of squamous epithelium similar to surface epidermis. There is a granular layer and the contents are formed by layered keratin. Usually the cyst wall is smooth, without rete ridges or epithelial proliferation, and cutaneous adnexae are absent in these cysts (Fig. 16.1). If the cyst is ruptured, a severe foreign body type inflammatory reaction is formed with foreign body giant cells, lymphocytes, histiocytes, plasma cells and neutrophils. Fragments of the cyst wall are usually seen among the inflammatory cells. Sometimes the epithelium has disappeared and the clue to the diagnosis is the presence of keratin scales among the inflammation.

Trichilemmal (Pilar) Cysts

Clinical

Smooth, firm, mobile nodules most commonly found on the scalp. The cyst walls are thick and are easily removed intact. Trichilemal cysts do not have a punctum, and inflammation is uncommon. They are multiple in 70 percent of cases.

Dermatopathology, edited by Ramón L. Sánchez and Sharon S. Raimer. ©2001 Landes Bioscience.

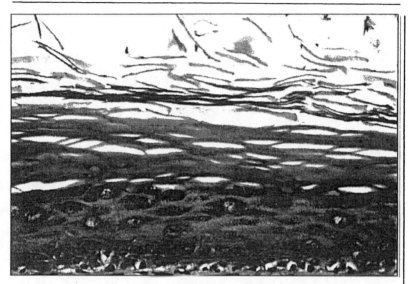

Fig 16.1. **Epidermal inclusion cyst**: The wall of the cyst shows keratinizing epithelium with a granular layer and the cavity contains layered keratin.

16

Fig 16.2. **Pilar cyst**: The wall of the cyst shows trichilemmal keratinization without a granular layer but rather pale, glycogen-containing keratinocytes.

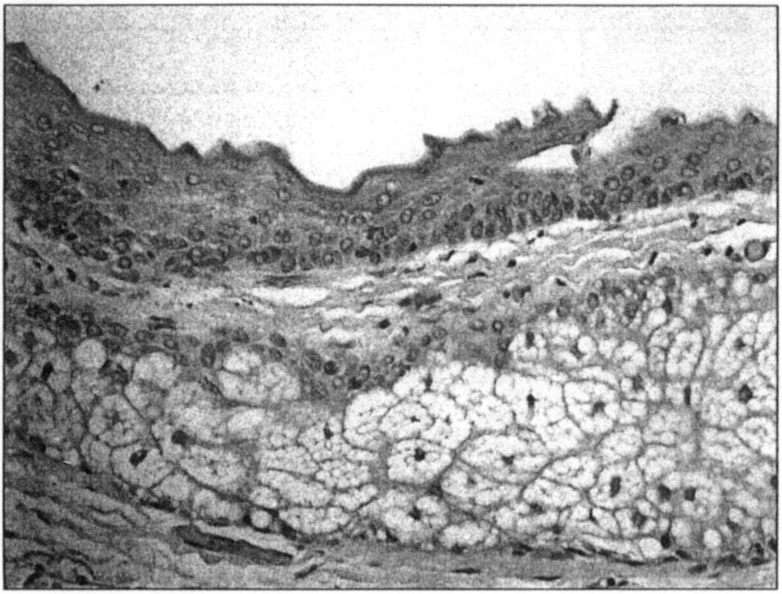

Fig. 16.3. **Steatocystoma**: The wall of the cyst contains sebaceous glands. A corrugated stratum lucidum separates the epithelium from the keratin contents.

Histology

The lining recapitulates the epithelium of the inner root sheath of the hair, maturing without forming a granular layer (trichilemmal keratinization). There is an abrupt keratinization from cuboidal, clear epithelial cells into dense, compact eosinophilic-staining keratin. Cholesterol clefts are common and calcifications also occur (Fig. 16.2).

Differential Diagnosis

Although the proliferating trichilemmal tumor also shows trichilemmal keratinization, it is essentially a solid tumor-like proliferation instead of a cyst.

Steatocystoma

Clinical

Steatocystoma is an autosomal dominant condition which presents as large as 1-3 cm, moderately firm cystic nodules adherent to the overlying skin which on puncture discharge a yellowish oily material. Found commonly in the axilla, sternal region and on the arms.

Histology

Cysts lined by a few to several layers of squamous epithelial cells without a granular layer. The luminal side of the epithelium is lined by a thick, wavy homogenous eosinophilic cuticular layer. Most lesions have sebaceous glands and/or abortive hair follicles in the cyst wall (Fig. 16.3).

Dermoid Cyst

Clinical

One to four centimeter subcutaneous cysts present at birth. Most commonly lateral to the upper eyelid. They are formed by the sequestration of cutaneous tissues along embryonal lines of closure.

Histology

Dermoid cysts are unilocular subcutaneous cysts lined by an epidermis with mature epidermal appendages; hair follicles and sebaceous glands are most common. Smooth muscle is often present, but bone and cartilage are not.

Vellus Hair Cyst (Eruptive)

Clinical

They are one to two millimeter asymptomatic follicular papules, most commonly on the chest of children and young adults.

Histology

It presents as a mid dermal cyst lined by squamous epithelium. The cyst contains laminated keratin and many vellus hairs.

Other Cysts

Bronchogenic cysts are rare, small, solitary lesions located above the sternal notch. They are buds that separate from the foregut during development of the respiratory tree. The cysts are lined by pseudostratified columnar epithelium. Cilia, goblet cells, smooth muscle and mucous glands may be present in the epithelium.

Cutaneous ciliated cysts are rare deep-dermal or subcutaneous cysts found on the lower extremities in young females. They are several centimeters in diameter and are usually multi-locular, filled with clear or amber fluid. Histology: The lining is cuboidal to columnar ciliated epithelium on intraluminal papillary projections. No mucin secreting cells are present.

Median raphe cyst of the penis is a small cyst present on the ventral penis, most commonly on the glans, in a young male. Histology: This cyst is lined with pseudostratified epithelium. Some epithelial cells have clear cytoplasm, but mucin-containing cells are uncommon.

Branchial cleft cysts are present as a swelling near the angle of the jaw. The cyst is lined by stratified squamous or pseudostratified ciliated epithelium. Lymphoid tissue with germinal centers is found in the wall.

Thyroglossal duct cyst is a congenital anomaly representing a remnant of the thyroid gland precursor. Presents as a fluctuant swelling up to three centimeters in the midline of the anterior neck. Histology: The cyst is lined by cuboidal or columnar

16

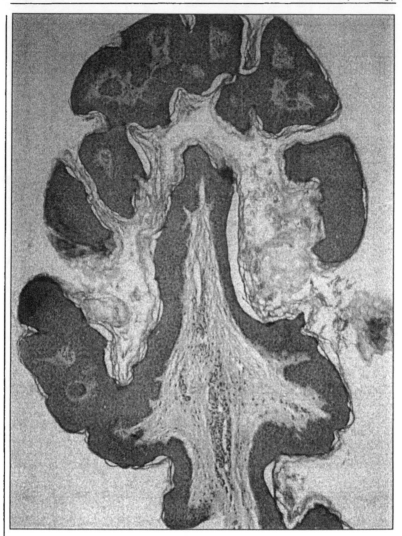

Fig. 16.4. **Achrochordon**: A squamous papilloma with fibrovascular stalk and irregular surface. Absence of cutaneous adnexae in the corium.

epithelium. Adjacent tissue may contain thyroid follicles, mucous glands and a heavy lymphocytic infiltrate. Smooth muscle, as is seen in bronchogenic cysts, is not present.

Pseudocyst of the ear is an asymptomatic one centimeter fluctuant swelling in the upper anterior pinna. If punctured, clear, yellow viscous fluid is released. The cyst wall is degenerated cartilage without an epithelial lining.

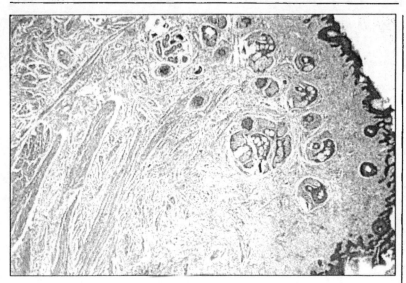

Fig 16.5. **Striated muscle hamartoma**: A papilloma with cutaneous adnexae and bundles of striated muscle spreading from the center toward the surface.

Section 44. Papillomas/Polyps

Achrochordons (Skin Tags)

Clinical

Achrocordons are soft, pedunculated fibromas that can occur as a solitary, large, bag-like growth or many one to several millimeter papules around the neck in the axilla or on the inner thighs.

Histology

Histology reveals a flat epidermis covering an exophytic lesion of loose collagen and fat. On the neck and axilla papillomatosis, hyperkeratosis and acanthosis are often present (Fig. 16.4).

Lipofibroma, a variant of achrochordon, is a term used when a significant portion of an achrochordon is made of fat cells.

Accessory tragus is a tag of skin found in the preauricular area. Histology shows numerous vellus hair follicles and often cartilage.

Striated Muscle Hamartoma

Clinical

Striated muscle hamartomas are rare solitary or multiple skin tags on the face or anterior chest in the newborn.

16

Table 16.1. Other hamartomas

Nevus comedonicus
Linear epidermal nevus
Nevus sebaceous
Accessory tragus
Striated muscle hamartoma
Connective tissue nevus
Nasal glioma

Histology

Histologically, a central core with individual fibers and bundles of skeletal muscle can be seen. Other mesenchymal elements can also be found in the lesion also (Fig. 16.5).

Section 45. Hamartomas

Accessory Nipple

Clinical

It is a light brown papule, often associated with hairs that appears in the embryonic milk lines.

Histology

A corrugated pattern of epidermis is present over a central pilosebaceous structure. The dermis contains smooth muscle and mammary glands.

Becker's Nevus

Clinical

It is a large brown macule arising in the second decade, mostly in men. Occurs most frequently on the chest, shoulder or upper arms and characteristically shows hypertrichosis.

Histology

The epidermis can have acanthosis and elongation of the rete ridges. Hyperpigmentation of the basal layer, melanophages in the upper dermis and an increased number of melanocytes are present.

Stain

Dopa-oxidase activity reveals increased melanocytes in the lesional skin.

Smooth Muscle Hamartoma

Clinical

It is an uncommon congenital or childhood lesion presenting as a hyperpigmented plaque or patch on the trunk, buttocks or proximal limbs. Perifollicular papules and

16

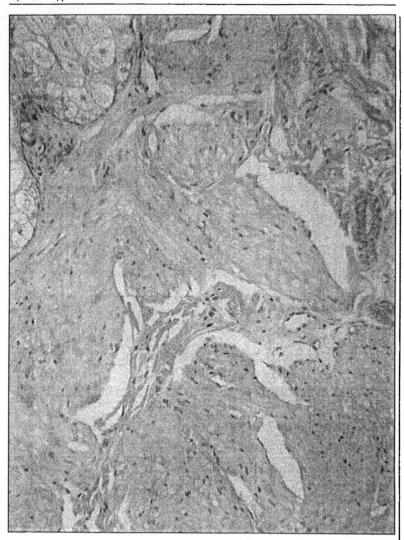

Fig 16.6. **Smooth muscle hamartoma**: Haphazardly arranged bundles of smooth muscle are seen criss-crossing the dermis.

hypertrichosis can be present. Pseudodarier's sign, when the lesion becomes raised with firm stroking, rarely can be elicited.

Histology

Dermis: Interlacing bundles of mature smooth muscle. The epidermis is often slightly hyperplastic and hyperpigmented (Fig. 16.6).

Stain

Trichrome will stain smooth muscle red.

Differential Diagnosis

Piloleiomyoma—here smooth muscle bundles in the dermis form a large aggregate.

Nevus Lipomatosus Superficialis

Clinical

An uncommon lesion with groups of soft, yellow to flesh colored papules and nodules presenting on the hip or buttock of a newborn or infant.

Histology

Clusters of adipocytes are found in the dermis. The proportion of fatty tissue varies from 10-50% of the dermis.

Differential Diagnosis

Focal dermal hypoplasia differs by its extreme attenuation of the collagen.

Section 46. Ectopic Tissues

Endometriosis

It is a bluish, often painful nodule found in the umbilicus or developing in the scar of a female following abdominal surgery. It can swell and bleed during menstruation. Endometrial glands and stroma in the dermis must be present for diagnosis. Menstrual bleeding can lead to hemosiderin deposits.

Endosalpingiosis

Endosalpingiosis consists of multiple brown papules seen in the umbilicus after salpingectomy. Each papule represents a unilocular cyst with papillary projections in the lumen. The lining consists of columnar cells, some secretory and others ciliated.

Umbilical Omphalomesenteric Duct Polyp

It is an eroded periumbilical lesion that shows ectopic gastric, small intestine, or colonic mucosa with a diffuse mononuclear infiltrate

Colostomy Site

After repair of a colostomy site the specimen received includes skin in continuity with the colonic wall and different amounts of fibrosis and of acute and chronic inflammation.

16

CHAPTER 17

Tumors of the Epidermis and of the Hair Follicles

Ramón L. Sánchez and Sharon S. Raimer

Section 47. Epidermal Tumors

Most people after their fourth decade have one or several epidermal tumors discussed in this section, particularly seborrheic keratoses. Fair skinned elderly individuals, with sun exposure in the past, often have one or more actinic keratoses in sun exposed areas.

Seborrheic Keratosis (SK)

Clinical

These are single or multiple discrete papules and plaques that can be seen anywhere, but they are especially common on the scalp and face, trunk, particularly on the back and to a lesser degree on the extremities. The typical seborrheic keratosis (SK) is a papule measuring between 0.5 and 1 centimeter in diameter that is stuck on the surface of the skin, and is covered by a greasy scale with a somewhat irregular surface. They are often pigmented, and may peel off on occasion but return. Sometimes SKs may appear polypoid resembling acrochordons. Flat hyperpigmented seborrheic keratoses are clinically difficult to distinguish from lentigo or nevi and some pigmented SKs difficult to differentiate from melanomas. That is especially true when the keratosis is irritated, in which case they often present a red surrounding area. Some patients have multiple depigmented seborrheic keratoses on the legs, and they are referred to as **stucco keratoses**. Another variation is **dermatosis papulosa nigra** that is very frequently seen on the face as multiple dark, small papules in blacks and other patients with darkly pigmented skin. The sign of Leser-Trelat (the existence of which has been questioned) refers to the sudden appearance of multiple seborrheic keratoses in individuals throughout their bodies which is associated with the development of visceral cancer.

Histology

The typical SK is an exophytic lesion that can be traced above a hypothetical line adjoining adjacent normal epidermis. It is characterized by hyperkeratosis, papillomatosis, with formation of pseudohorn cysts and acanthosis. The latter refers more to proliferation of epidermal buds in papular-like formations, rather than pure thickness of the epidermis. There are also anastomosing bands of epithelia. The cells

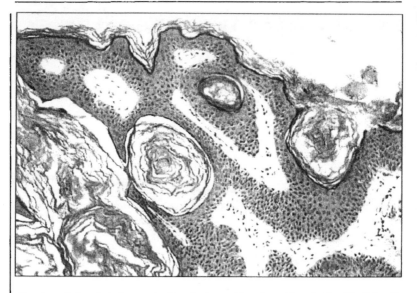

Fig 17.1 . **Seborrheic keratosis**: Exophytic proliferation of basaloid epithelial cells with anastomosing bands and pseudo-horn cysts.

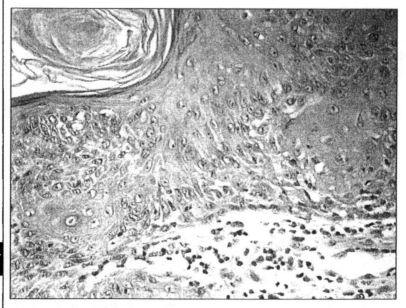

Fig 17.2. **Irritated seborrheic keratosis**: Whorled collections of mature keratinizing epithelial cells (squamous eddies) and a lymphocytic infiltrate in the dermis.

17

composing seborrheic keratoses are for the most part basaloid, homogeneous, with regular nuclei. They usually display intercellular bridges, and they are seen in the whole thickness of the epidermis (i.e., they are not limited to certain layers of the epidermis). The dermis between the epithelial proliferation shows no marked fibrosis (Fig. 17.1).

There are several variations of seborrheic keratosis that should be recognized. Some SKs show rather mature epidermal cells rather than the typical basaloid appearance. Some of these lesions have hyperkeratosis, and they have been termed **hyperkeratotic seborrheic keratoses**. In these cases the diagnostic feature is the papillomatous architecture. Other seborrheic keratoses are rather flat with minimal acanthosis, although showing typical basaloid cells. Among the flat keratoses there is a subtype called **reticulate seborrheic keratosis**, in which thin strands of basaloid cells fall down into the papillary dermis from a thin epidermis. **Irritated seborrheic keratosis** (Fig. 17.2) is characterized histologically by the presence of squamous eddies among the basaloid cells of an SK. Those lesions may not show any associated inflammation. However, SKs are often seen in association with a considerable amount of inflammation, and should be diagnosed as inflamed or irritated seborrheic keratoses. Irritated seborrheic keratosis is an important diagnosis, since clinically they are difficult to differentiate from squamous cell carcinoma. **Pigmented seborrheic keratosis** shows similar histology as the preceding described lesions, but in addition it has associated melanin pigment and occasionally melanocytes. **Melanoacanthoma** refers to an SK with abundant large entoptic melanocytes among the tumor cells with abundant melanin deposition. **Clonal seborrheic keratosis** shows well-defined epithelial nests of differentiated cells, sometimes clear, similar to the effect seen in Bowen's and referred to as the Borst-Jadassohn effect. Occasionally we come across lesions that have the architecture of seborrheic keratosis but display marked severe atypia with abundant mitosis, such as seen in Bowen's disease. If the dysplastic changes warrant it, we classify these lesions as Bowen's disease, although in the past we referred to them as Bowenoid keratosis.

Differential Diagnosis

Hyperkeratotic seborrheic keratosis cannot be differentiated on histologic grounds from other papillomatous processes such as linear epidermal nevi, nevus sebaceous and acanthosis nigricans. Reticulated SKs can resemble fibroepithelioma of Pinkus. The correct diagnosis is established by the absence of fibrosis in the dermis in seborrheic keratosis and full epidermal thickness of involvement.

Clear Cell Acanthoma

Clinical

Clear cell acanthoma presents as solitary lesions frequently on the anterior surface of legs or thighs, they present as reddish to tan colored papules with well-defined borders.

17

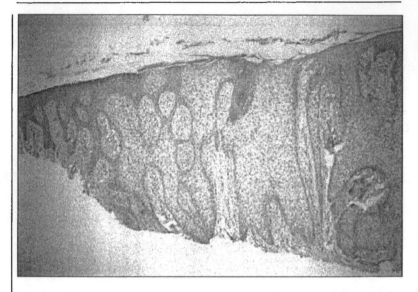

Fig 17.3. **Clear cell acanthoma**: There is acanthosis and clear change of the keratinocytes (glycogen rich). Also, there is exocytosis of neutrophils and subcorneal microabscesses.

Histology

The epidermis is irregularly acanthotic and shows a sharp demarcation between the area of hyperplasia and the normal epidermis. The individual keratinocytes show clear abundant cytoplasm which contrast with the cytoplasm of the keratinocytes in normal epidermis. Although the changes in the keratinocytes are continuous throughout the lesion, occasional acrosyringium (intraepidermal eccrine ducts) can be visualized. Epithelia of the acrosyringium is not involved by the clear cell changes, and therefore appears eosinophilic, such as in the normal epidermis (Fig. 17.3). There are neutrophils that extend into the epidermis. The stratum corneum shows parakeratosis and occasional neutrophilic microabscesses.

Occasional lesions may show multiple dendritic melanocytes in the epidermis. In this circumstance the separation between melanoacanthoma and clear cell acanthoma is blurred.

Special Stains

PAS will show an increased amount of glycogen in the tumor cells.

17

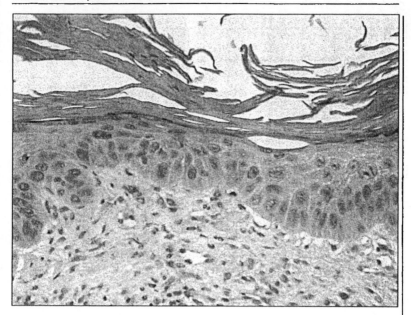

Fig 17.4. **Actinic keratosis**: Parakeratosis overlying a thin epidermis in which there is dysplasia of basal keratinocytes with formation of buds of atypical epithelium.

Premalignant, Malignant Conditions

Actinic (Solar) Keratosis

Clinical

Actinic keratosis are hyperkeratotic papules seen on sun exposed areas, particularly the face and forehead, ears, neck, and dorsal hands and forearms. They are often small with a red to yellowish color, they are sometimes hyperpigmented and they feel rough to the touch. Occasionally they can be hyperkeratotic, presenting as a **cutaneous horn**, particularly on the forearms. Since they are related to sun exposure, and radiation exposure is cumulative, they are seen more often after the fifth decade and most commonly in fair skinned individuals who have had excessive sun exposure. They are usually multiple and may precede the development of squamous cell carcinoma. They were first described at the end of the nineteenth century on the skin of farmers and sailors.

Histology

Histologic changes in actinic keratoses are seen at the level of stratum corneum, stratum of Malpighii, and in the dermis. Characteristically there is parakeratosis overlying an actinic keratosis. It is in the stratum of Malpighii where diagnostic changes are found, and they are characterized by dysplastic features and architectural disarrangement of the lower layers of the epithelia. Individual cells show larger hyperchromatic nuclei of different sizes (anisocytosis). Dyskeratosis (individual cell keratinization) is characteristically seen. There is often a tongue-like proliferation of epithelia, or small nests of dysplastic keratinocytes extending into the papillary dermis (Fig. 17.4). Occasionally the dysplastic changes described for the two or three lower layers of the stratum of Malpighii involve the full thickness of the epidermis, resembling Bowen's disease. Those lesions are termed **Bowenoid actinic keratoses**. The third component is solar elastosis involving the upper dermis, which is constantly found in actinic keratosis. Pigmented actinic keratosis show, in addition to the changes described above, increased melanin pigmentation. These lesions can be difficult to differentiate from lentigo solaris, particularly if the dysplastic changes are not prominent.

Differential Diagnosis

Bowen's disease can be difficult to differentiate from Bowenoid actinic keratosis, although the former usually also involves hair follicles. Some dermatopathologists use size of the lesion as a distinguishing point, with actinic keratosis being less than one centimeter in maximum diameter. It is often difficult to differentiate an advanced actinic keratosis from an in-situ squamous cell carcinoma. The distinction is often arbitrary. However, the distinction can be meaningful since squamous cell carcinomas in-situ are usually treated more aggressively. **Hypertrophic actinic keratoses** are seen often on the dorsum of the hands and forearms, and occasionally may become cutaneous horns. Histologically they are characterized by orthokeratosis and papillomatosis with acanthosis. The lower layers show dysplastic changes. Hypertrophic actinic keratoses are often incompletely shaved, and the biopsy may lack the diagnostic lower layers. **Atrophic actinic keratoses** show a very diminished stratum of Malpighii, which is often flat with disappearance of the rete regions and a dysplastic basal layer, which is sometimes artifactually separated from the rest of the epithelium (acantholysis). **Actinic chelitis** refers to solar-related damage resembling actinic keratoses on the lower lip. It can be quite extensive and is clinically characterized by a whitish hyperkeratotic plaque. Histologically it shows parakeratosis and acanthosis of the epithelia with dysplastic changes. The distinction between actinic chelitis and early squamous cell carcinoma is very important for the lip since carcinomas in this location have a more aggressive behavior than in other sun exposed areas.

Benign Lichenoid Keratosis/Lichenoid Actinic Keratosis

Clinical

They are hyperkeratotic papules, usually single, seen on the upper trunk or the arms.

Histology

Histologically they classically show parakeratosis or orthokeratosis, irregular acanthosis of the epidermis with a saw-toothed appearance and a lichenoid lymphocytic infiltrate along the papillary dermis, which reaches the basal layer of the epithelia. Eosinophilic, colloid bodies are seen in the papillary dermis as well as in the keratotic cells in the epithelia. The histologic findings are very similar to those of lichen planus from which it is separated by the presence of parakeratosis (Fig. 17.6). The clinical

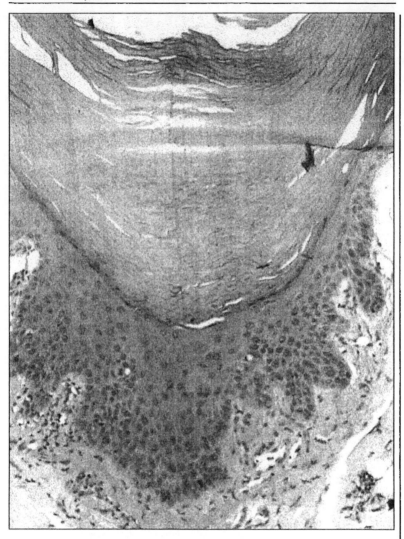

Fig 17.5. **Hypertrophic actinic keratosis**: A thickened compact orthokeratosis, overlying an acanthotic epidermis with dysplastic features of the basal keratinocytes.

history of a single papule is also helpful in distinguishing these two entities. Two histologic variations are the so-called lichenoid actinic keratosis and lichenoid seborrheic keratosis. They represent inflamed keratoses with preservation of the distinguishing features of these two entities.

17

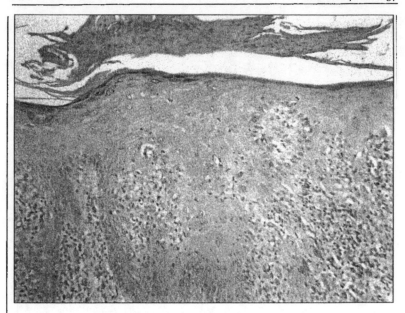

Fig 17.6. **Benign lichenoid keratosis:** Parakeratosis, acanthosis and a lichenoid and interface infiltrate with necrotic keratinocytes.

Clonal Intraepidermal Proliferations/Borst-Jadassohn Effect

Clinical
This entity refers to the so-called **intraepidermal epithelioma** or **intraepidermal carcinoma**, and since it can be seen in different entities, it does not have a specific clinical appearance.

Histology
There are well-defined, discrete nests of homogeneous appearing keratinocytes in the epidermis. The cells inside the nests are different from the surrounding keratinocytes, or epithelial cells, and are characterized by rounded to oval nuclei and scanty cytoplasms. They can be light and even clear, suggesting the possibility of metastatic adenocarcinoma. Special stains, however, fail to reveal mucin in the clear cytoplasm (Fig. 17.7).

Note: The above histologic findings can be seen in seborrheic keratosis, Bowen's disease, intraepidermal eccrine poroma. (hydroacanthoma simplex), and malignant eccrine poroma.

Bowen's Disease

Clinical
It is characterized by well-defined reddish plaques with whitish scales. It often is seen on the trunk and extremities, particularly the legs, and occasionally in sun

17

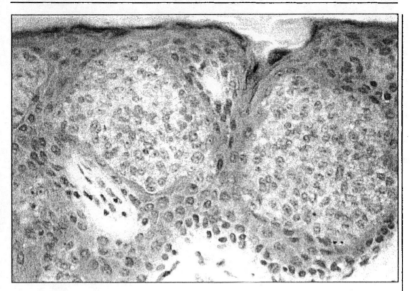

Fig 17.7. **Clonal Bowen's**: Well delimited nests of clear, atypical cells are seen in the epidermis, surrounded by normal keratinocytes.

protected areas. Clinically it should be differentiated from an inflammatory dermatosis, such as chronic dermatitis or psoriasis.

Histology

Bowen's disease is considered a carcinoma in-situ, although the severity of the histologic changes can be quite variable, from moderate dysplasia to severe or frank carcinoma. The histologic findings are lack of maturation of the keratinocytes, usually involving full thickness of the epidermis. The cells appear basaloid with a high nuclei-cytoplasmic ratio. At the same time there is no arrangement to the architecture, as opposed to the relatively orderly arrangement of the normal epidermis. In more advanced cases, there is marked anisokariosis (different nuclear sizes) with some of the nuclei becoming very enlarged, giant and hyperchromatic. Mitoses are a consistent finding, and they are often atypical. It is common to see tripolar mitosis on a clear cell background. By definition there is lack of invasion of the dermis (Fig. 17.8).

Differential Diagnosis

Bowen's disease should be differentiated from Bowenoid actinic keratosis. Helpful in differentiating between the two are the size of the lesion, with Bowen's disease often being at least one centimeter in diameter, and involvement of the hair follicles, i.e., Bowen's disease involves full thickness of the follicular epithelia, while actinic keratosis is seen involving only the basal layers of the most superficial part of the follicle. It is not uncommon to see lesions with the architecture of seborrheic keratosis, which show, however, dysplastic features which would qualify them as Bowen's disease.

17

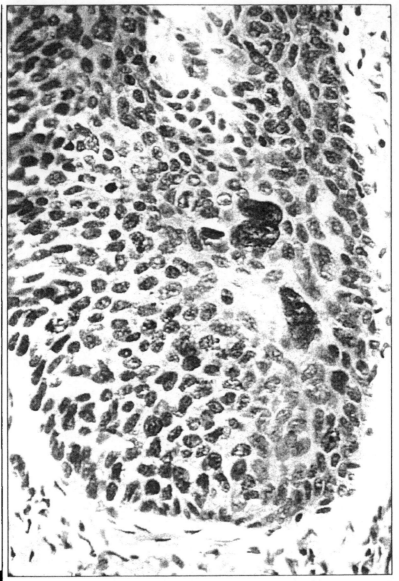

17

Fig 17.8. **Bowen's disease**: Full thickness atypia of the epidermis with lack of maturation, abundant mitoses, large atypical cells and architectural disarrangement.

If the dysplasia seen is beyond that of irritated seborrheic keratosis, and there is atypical mitosis, we prefer to classify these lesions as Bowen's disease.

Note: Although Bowen's disease is an in-situ carcinoma, it is known generally to remain intraepithelial for many years. Occasionally Bowen's disease becomes invasive, and if it does, it acquires the capability to metastasize. For that reason Bowen's is treated as a carcinoma. In reporting a lesion of Bowen's, to dissipate any doubts, we use the term squamous cell carcinoma in-situ, Bowen's type.

Squamous Cell Carcinoma (SCC)

Clinical

It is most often seen in sun exposed areas, particularly the head and neck, and the upper extremities, and less commonly on the upper trunk or legs. Lesions often present as a red plaque or nodule which is usually hyperkeratotic but may be ulcerated. Squamous cell carcinomas sometimes arise in areas of previous radiation therapy, in relation to old burn scars (Marjolin's ulcer), in chronic ulcerations or associated with chronic infections, such as osteomyelitis sinuses. Most sun-induced squamous cell carcinomas behave in a relatively, nonaggressive fashion, although occasional tumors can produce metastases. There are several exceptions to this rule, particularly with lesions on the ears and above the eyes, and on the dorsum of the hands, which often behave more aggressively. Squamous cell carcinomas arising on the lower lip, even though actinically induced, are quite aggressive and should be treated early by complete excisions. Squamous cell carcinomas arising in transformed skin, such as scars or chronic infections, also behave more aggressively. In the mucosa or skin of the external genitalia, squamous cell carcinoma may follow infection by papilloma viruses (HPV-16 and 18). Those tumors should also be considered aggressive tumors and should be treated early.

Histology

Classically SCCs have been histologically divided into well, moderately and poorly differentiated tumors. Well-differentiated SCCs resemble the cells in the spinous layer of the stratum of Malpighii, exhibiting large eosinophilic cytoplasm and round to oval nuclei. Intercellular bridges are frequently seen, and they often show squamous cells or horn cysts. At the periphery of tumor nests, cells that are less differentiated with a higher nucleus/cytoplasmic ratio can be seen. Mitoses are frequently found in these peripheral areas, including atypical mitoses. Moderately differentiated SCCs show cells with smaller cytoplasms and more prominent nuclear atypia. Dyskeratosis, hyperchromasia, anicytosis, and anisokariosis are present. Still the tumor can be easily recognized as a squamous cell carcinoma (Fig. 17.9). Poorly differentiated tumors can present with small atypical cells or larger cells without clear signs of keratinization. Occasionally poorly differentiated carcinomas may display spindle cells which may require special stains to diagnose as carcinoma, as opposed to melanoma or sarcoma (Fig 17.10). If the atypical squamous cells are still attached to the epidermis, a diagnosis of carcinoma in-situ is made. In such cases it is often difficult to separate a carcinoma from an actinic keratosis, and the distinction is made arbitrarily based on the degree and amount of dysplastic epithelia seen. In this context a

17

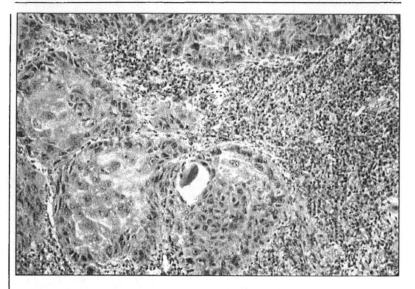

Fig 17.9. **Squamous cell carcinoma**: Nests of atypical squamous epithelial cells are seen invading the dermis surrounded by a marked lymphoplasmacytic infiltrate.

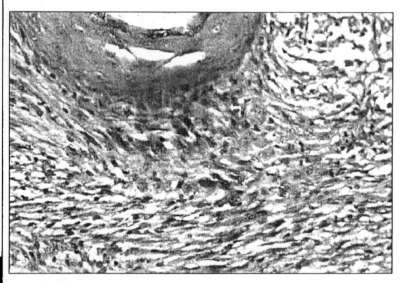

Fig 17.10. **Spindle cell squamous cell carcinoma**: An atypical, malignant spindle cell tumor connected to the epidermis. Stains are needed to rule out other spindle tumors.

pathologist should be reminded that a diagnosis of actinic keratosis will imply that no further treatment is necessary, while most clinicians will excise or curette a lesion in which the diagnosis of squamous cell carcinoma in-situ was made. There is often a prominent infiltrate around the tumor nodules, made up predominantely of lymphocytes. Plasma cells are also seen, sometimes in great numbers associated with SCCs on the face and near mucosa.

The biological behavior of individual tumors has been associated with the grade of differentiation, with well-differentiated tumors generally having a slower rate of progression (although they may occasionally behave very aggressively) than poorly differentiated carcinomas (Broder's). Other factors influencing prognosis are tumor necrosis and vascular or neural invasion and the immune status of the host. SCCs occasionally recur and rarely metastasize, therefore they should be treated as malignant tumors.

Other Related Entities

Necrotizing sialometaplasia occurs on the hard palate. It presents clinically as an ulcer. Histologically in addition to abundant inflammation, there is metaplasia of the minor salivary gland which can be easily confused with squamous cell carcinoma. **Arsenical keratoses** are characterized by hyperkeratotic papules, particularly on the palms, hands and forearms, many with the histologic features of Bowen's disease, related to chronic ingestion of arsenic. **Xeroderma pigmentosa**, an autosomal dominantly transmitted congenital disease characterized by the inability to repair DNA. It is characterized clinically by multiple keratosis and actinic damage beginning in early childhood and the development of numerous skin malignancies, including squamous cell carcinoma, melanoma, and basal cell carcinoma. **Lymphoepithelioma-like carcinoma of the skin** is a histologic variant of squamous cell carcinoma in which strands and nests of squamous cells are surrounded and embedded in a very prominent lymphocytic infiltrate which may obscure the presence of tumor cells. The lymphocytes are morphologically normal. Immunoperoxidase stains show clearly the differentiation between cytokeratin positive cells and lymphocytes.

Keratoacanthoma

Clinical

These are rapidly growing tumor nodules, particularly seen on the head, face and extremities with a central keratinous plug. It is said that these tumors involute spontaneously in weeks if left untreated.

Histology

The normal epidermis is lifted and shows a central invagination from which a cyst-like nodule of neoplastic epithelia arises. The center of the cyst shows epithelial papilla and abundant keratin. The most external wall of the cyst is irregular with epithelial bands which, because of the histologic sectioning, appear to be separated from the main tumor and invading the stroma. In general the neoplastic cells in keratoacanthoma are hypermature with large glassy, pink to clear cytoplasm (Fig. 17.11).

17

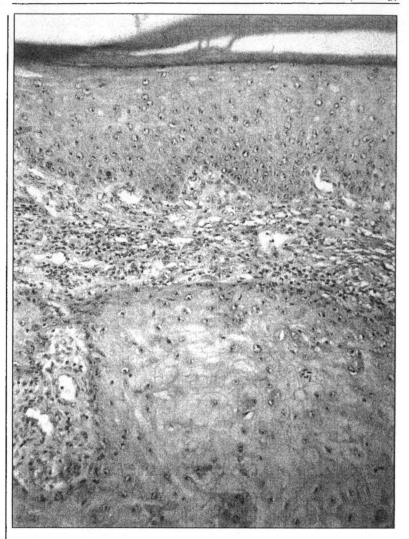

Fig 17.11. **Keratoacanthoma**: Nests of mature keratinizing epithelium are seen in the dermis still connected to the epidermis. A mixed infiltrate surrounds the tumor nests.

Mitoses are not uncommonly seen. Internalization of elastic fibers in the tumor cells is frequently seen. Interepithelial neutrophilic microabscesses are also a feature of keratoacanthoma. The distinction of keratoacanthoma from a well-differentiated squamous cell carcinoma is not possible when the lesion has not been completely excised. Even with a complete excision, many dermatopathologists prefer to classify these lesions as well-differentiated squamous cell carcinomas, keratoacanthoma type.

Verrucous Carcinoma

Clinical

Verrucous carcinoma manifests in a papillomatous verrucous surface, sometimes associated with papilloma viruses, such as condyloma or laryngeal papillomatosis. It is also seen on the extremities, particularly toes and fingers.

Histology

Resembles a verruca, possibly with more extensive acanthosis. The deeper margin of the tumor shows rounded pushing borders without definitive invasion of the dermis. These tumors often lack convincing atypia to help in the diagnosis of carcinoma, but they are recurrent lesions that can eventually become invasive.

Basal Cell Carcinoma

Clinical

Basal cell carcinoma is the most common malignant tumor, often seen on the face, forehead or upper trunk, although it can be seen elsewhere. It is not present on mucosa. The typical clinical presentation is that of a papule with pearly gray color and overlying telangiectasia. If the tumor is long standing, it may present as an ulcer, often with a rolled border. Morpheaform BCC's present as an atrophic or sclerotic appearing patch or plaque. Superficial basal cell may present as a reddish patch. Basal cell nevus syndrome is an autosomal dominantly inherited disorder in which patients develop multiple basal cell carcinomas early in life, associated with other abnormalities, including palmar pits, jaw cysts, hypertelorism, frontal bossing and other bone abnormalities such as bifid ribs.

Histology

Classic basal cell carcinoma is characterized by nests of basaloid cells displaying hyperchromatic nuclei and a scanty cytoplasm. These nests are connected to the epidermis or infiltrate the dermis and occasionally the subcutaneous and underlying tissues. The tumor nests show peripheral palisading of cells. Tumor cells are for the most part homogeneous without displaying atypical features or abundant mitosis. Between the tumor cells there is often mucinous material that separates the cells and occasionally produces pseudoglandular spaces. Occasionally these spaces are so prominent that it resembles an adenoma rather than a carcinoma (glandular BCC). Surrounding the tumor nests there is a prominent reaction of the stroma characterized by the deposition of new collagen by young fibroblasts and abundant mucous substances. Routinely processed basal cells display a characteristic artifact consisting of a cleft caused by detachment of the tumor nest from the surrounding stroma (Fig. 17.12). This clear space is characteristic of BCCs, and occasionally its presence helps in the diagnosis. Vascular or lymphatic invasion by BCCs is seldom seen.

Subtypes

Fibroepithelioma of Pinkus is a basal cell carcinoma in-situ, characterized by thin strands of basaloid epithelial cells projecting from the epidermis with frequent

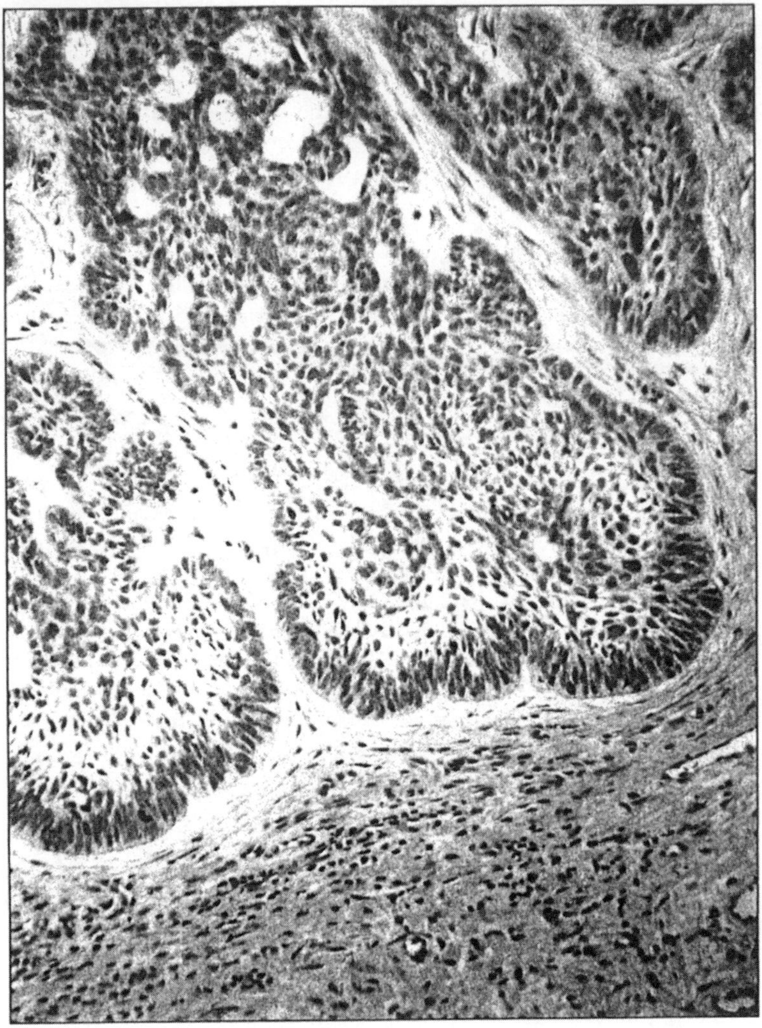

Fig 17.12. **Basal cell carcinoma**: Nests of basaloid cells with peripheral palisading surrounded by a loosely fibrotic stroma often with detachment artifact.

anastomosis. The stroma is often fibrous, and this feature plus the presence of adjacent classical BCC helps distinguish this entity from a reticulate seborrheic keratosis. **Morpheaform BCC** is an infiltrating BCC with a marked fibrous reaction of the stroma. The tumor cells are arranged in strands, one or two cells thick, and they often display atypical nuclear features (Fig. 17.13). This type of tumor appears to be more aggressive than classic BCCs and should be excised with adequate margins. **Pigmented BCCs** show solid nests of basaloid cells, many of which contain melanin. The pigment is also seen between tumor cells and in melanophages. **Micronodular BCC** is characterized by small nests, often with little stromal reaction. This is an aggressive tumor which often is seen infiltrating deeply in the soft tissue. **Syringomatous** or **eccrine-like BCCs** show areas of eccrine differentiation. The tumor cells in basal cell nevus syndrome display the morphology of regular basal cell carcinomas. Some of these tumors are occasionally calcified. **Baso-squamous carcinoma** shows features of both BCC and SCC, and they may be difficult to classify under only one heading. These tumors, in general, behave more aggressively than classic BCC, and, in fact, account for most of the few cases recorded in literature of basal cells with lymphatic or hematogenous metastasis.

Paget's Disease: Mammary and Extramammary

Clinical

Mammary Paget's disease is characterized by a slightly indurated, eczematous rash on the nipple, often unilateral, that does not resolve after conventional topical steroid therapy. Extramammary Paget's disease presents in the perineal, perianal and external genital area as eczematous patches with geographic borders.

Histology

Both types are characterized by the presence of clear, atypical neoplastic cells embedded in the epidermis without an underlying apparent carcinoma. For the most part the neoplastic cells are separated from each other, although small nests or glands can be seen, particularly in the extramammary variant. They are often located in a basal position, although characteristically separated from the papillary dermis by a layer of epithelial basal cells (Fig. 17.14). Paget's disease of the nipple implies the presence of an underlying ductal carcinoma, while only about 10 to 15 percent of cases of extramammary Paget's disease are associated with an underlying malignancy, often from the rectum.

Special Stains

Paget's cells often show mucosubstances in the cytoplasm as demonstrated with mucicarmin, PAS, and other mucin stains. Mammary Paget's cells usually stain positive for EMA, CAM 5.2, and GCFDP 15 (BRST2) with immunoperoxidase stains. CEA and cytokeratine cocktail stains are also usually positive, while S-100 is negative. CK7 and CK20 are also cytokeratins used in the diagnosis of Paget's disease.

Differential Diagnosis

This includes Pagetoid melanoma and Bowen's disease.

Merkel Cell Carcinoma

Clinical

These are nodules often on the head and neck region, but also elsewhere, most commonly seen in elderly patients.

17

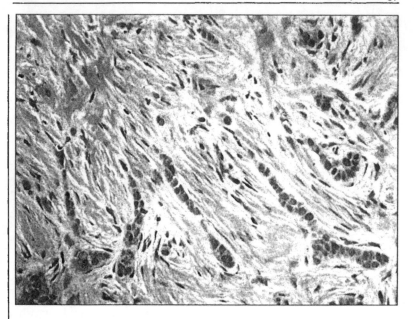

Fig 17.13. **Morpheaform basal cell carcinoma**: One or two cells thick cords of basaloid cells are infiltrating a densely fibrotic stroma.

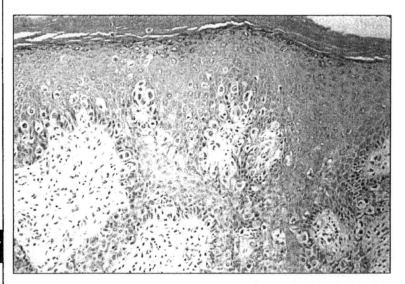

Fig 17.14. **Paget's disease**: Individual cells or small groups of clear appearing tumor cells are invading the epidermis generally above the basal layer.

Histology

Merkel cell tumors, also called trabecular carcinomas, are characterized by strands or trabeculae, as well as nests of undifferentiated tumor cell nests also present in the dermis. The tumor is usually not continuous with the epidermis, although occasionally there is a coexistent squamous cell carcinoma which can be present in situ or invasive in continuity with the undifferentiated carcinoma. The tumor cells show a round to oval nucleus with fine chromatin and only a small or inconspicuous nucleolus. The cytoplasm is scant and purplish. The tumor cells are very similar to those of small cell undifferentiated carcinoma of the lung (Fig. 17.15).

Special Stains

Merkel cells stain for cytokeratin with CK20 and Cam 5.2, usually displaying a paranuclear dot. Some tumors are positive for neurofilaments, and most of them are positive for NSE.

Differential Diagnosis

Includes undifferentiated carcinoma of the lung and lymphoma. Merkel cell carcinomas cannot be properly separated from the former by H&E sections although the presence of a paranuclear dot with cytokeratin or the presence of neurofilaments are strong indicators of Merkel cell carcinoma. Lymphoma or leukemia can be distinguished by the cytologic features and the positivity of lymphoid markers with immunoperoxidase stains.

Section 48. Hair Follicle Tumors

Pilar Sheath Acanthoma/Dilated Pore of Winer

Clinical

Dilated pore of Winer may present on the trunk and face, while pilar sheath acanthoma occurs almost exclusively on the face. They both appear as a large comedo.

Histology

It is a large dilated cystic follicle filled with keratin and lined with squamous epithelium. The basal layer in both entities is irregular with epithelial ridges, which is a distinguishing feature from an epidermal inclusion cyst in which the lining epithelium, as well as the basal layer appear flat. In pilar sheath acanthoma small vellus hairs are seen in the wall of the cyst.

Inverted Follicular Keratosis

Clinical

Inverted follicular keratosis appears as a cyst without any specific site predilection.

17

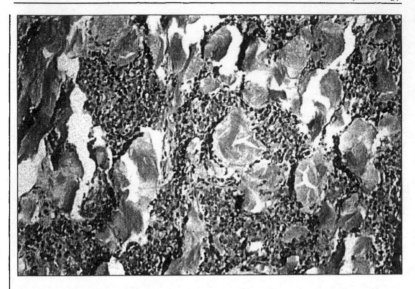

Fig 17.15. **Merkel cell carcinoma:** Trabeculae and groups of undifferentiated basaloid cells with scanty cytoplasm and abundant mitoses are seen invading the dermis.

Histology

A cystic follicle shows an epidermal lining that resembles that of an irritated seborrheic keratosis. The cells are small, basaloid and exhibit squamous eddies. Inverted follicular keratosis has been considered identical to an irritated seborrheic keratosis, however, the architecture is completely different and is a distinguishing feature.

Pilomatricoma (Calcifying Epithelioma of Malherbe)

Clinical

Pilomatricoma presents in children and young adults predominantly, although it can be seen in all ages. It has a predilection for the head, shoulders and upper extremities, and manifests as a cystic lesion that is very firm, producing the so-called tent sign.

Histology

Pilomatricomas are seldom removed intact and therefore appear as fragmented or ruptured cysts. Histologically there are four components which are almost always seen. The live epithelium is characterized by small, basaloid cells with high N/C ratio which resemble the cells in the hair follicle bulb. There are also fragments of epithelium with ischemic necrosis, characterized by cells without any distinct differentiation between the nucleus and the cytoplasm, both of them appearing light pink. These cells have been termed ghost cells. Both types of epithelium, the live and the ischemic epithelium, are diagnostic of pilomatricoma. In addition there are areas of calcification and areas of necrosis. Mitotic figures or atypia are infrequent. A marked foreign body reaction is also present (Fig. 17.16).

17

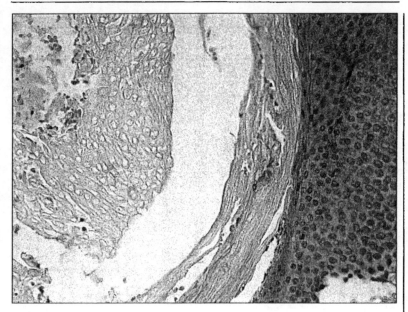

Fig 17.16. **Pilomatricoma**: Ghost cells are seen on the left, opposite to basaloid epithelial cells on the right. Missing are calcification and a foreign body giant cell reaction.

Trichoepithelioma

Clinical

Solitary trichoepithelioma occur usually on the face, often on or near the nose, without any specific age predilection. Multiple trichoepithelioma syndrome, also known as Brook's syndrome, is characterized by the presence of multiple papules, particularly on the head and neck, with a predilection for developing around the nose and mouth often in the first or second decade. Other areas also may be involved, particularly perineal areas. The papules in Brook's syndrome reach a size of a few millimeters, and then they usually persist indefinitely. Cylindromas, particularly on the skull, are often part of this syndrome. Malignant tumors, however, are usually not seen.

Histology

Both solitary and multiple trichoepitheliomas exhibit similar histologic features. There is an epithelial proliferation in the dermis which is characterized by the presence of a small epithelial cyst filled with keratin. From the wall of this small cyst there is a proliferation of basaloid cells that extends peripherally in a reticulated pattern and adjoins other cysts. Solid areas of basaloid cells are also seen as well as cells with pink cytoplasmic granules resembling trichohyalin granules. It is helpful to see these tumors on low magnification to observe a lobular architectural arrangement. The stroma surrounding the tumor cells appears somewhat fibrotic but there is no retraction artifact as is seen in basal cell carcinomas (Fig. 17.17).

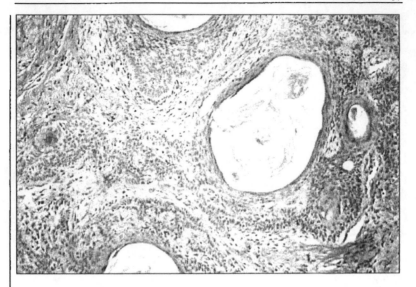

Fig 17.17. **Trichoepithelioma**: Lobular arranged epidermoid cysts connected by basaloid strands to immature hair follicles and hair bulbs.

Comment

Trichoepitheliomas are considered to be hamartomas, without malignant potential, and are treated as such. Histologically they resemble keratotic basal cell carcinomas, from which they are often difficult to differentiate. In the case of a solitary trichoepithelioma, the distinction from basal cell carcinoma may be almost impossible, but is important because of the different behavior as well as the different treatment modes used for each entity.

Desmoplastic Trichoepithelioma

Clinical

They occur on the face, particularly on the upper lip or near the nasal-labial fold, often in women. Lesions usually present as an atrophic plaque which resembles morphea-like basal cell carcinoma.

Histology

Histologically they show strands of tumor cells and small cysts surrounded by fibrous stroma. These tumors can be difficult to differentiate from microcystic adnexal carcinoma. However, the former is a benign superficially located tumor without infiltrative borders, while the latter is an infiltrating tumor, usually seen throughout the dermis, and often in the subcutaneous tissue, which frequently has perineal invasion. Calcifications and absence of deep dermal involvement favor trichoepithelioma.

17

Trichofolliculoma

Clinical

They occur on the face where they may show a tuft of small vellus hairs protruding from a single follicular opening.

Histology

The architecture of the lesion is that of a dilated hair follicle, from which numerous buds of follicular epithelium emerge and project into the surrounding dermis. Some of the epithelial buds are fully formed vellus hairs which penetrate into and through the follicular opening.

Trichilemmoma

Clinical

Solitary trichilemmoma occur primarily on the face where they can be confused with a wart or seborrheic keratosis. Multiple trichilemmomas, as seen in Cowden's syndrome, an autosomal dominantly transmitted disease, are seen predominantly on the face, but also on the chest and extremities. Other components of the syndrome include squamous papillomas, some of which resemble trichilemmomas on the lips and other mucosa, other skin tumors including keratoacanthomas and a high incidence of visceral malignancies from primary sites such as the thyroid, breasts and GI tract.

Histology

There is a bulb-like epithelial proliferation that projects into the papillary dermis. The epithelium expands in the dermis, and there is a clear appearance of the cells, which is due to the presence of a glycogen in the cytoplasm of the keratinocytes, as demonstrated with PAS stain. There is a palisading arrangement of the cells at the periphery of the bulbar expansions of the epidermis. Beneath the epithelium is a thick eosinophilic and homogeneous basal membrane (Fig. 17.18). The histologic findings in multiple trichilemmomas are identical to those in solitary trichilemmomas.

Other Tumors of Hair Follicle Origin

Trichoblastomas represent follicular hamartomas with a more primitive differentiation than seen in trichoepitheliomas, being characterized by basaloid nests which need to be differentiated from basal cell carcinomas. **Tumor of the follicular infundibulum** exhibits a plate-like basaloid cell proliferation that runs parallel to the epidermis, and from which several vellus hairs emerge.

Fibrofolliculoma and **trichodiscoma** can be seen as solitary or multiple tumors and they represent a proliferation of hair follicle epithelium and of specialized mesenchyme similar to that of the papilla of the hair follicle. Mucin is an important component of trichodiscoma, a tumor composed almost exclusively of mixoid fibrous tissue. **Trichoadenoma** exhibits multiple keratin cysts throughout the dermis.

17

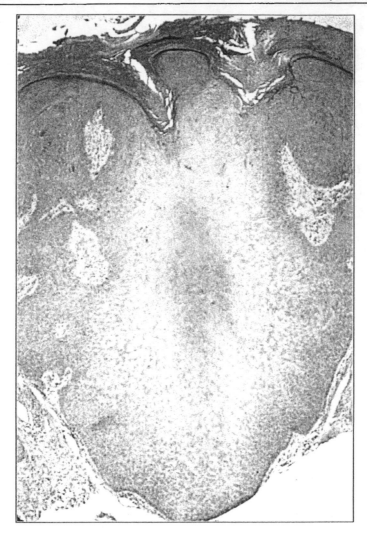

Fig 17.18. **Trichilemmoma**: A nodular epithelial proliferation of clear cells connected to the epidermis with peripheral pallisading and thickened basal membrane.

Proliferating Trichilemmal Tumor

Clinical

It is seen in elderly patients almost exclusively on the scalp, as solid tumors resembling pilar cysts.

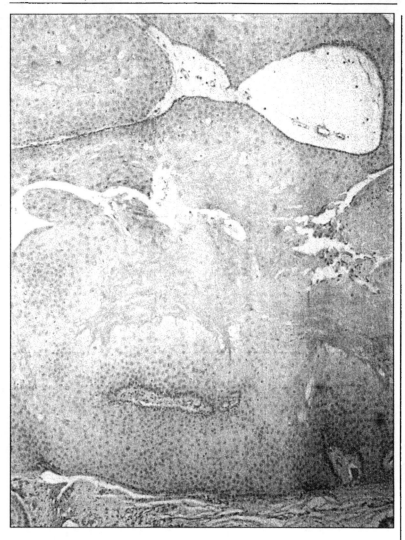

Fig 17.19. **Proliferating trichilemmal tumor**: A complex cystic tumor with nests of mature epidermal cells and trichilemmal keratinization.

Histology

17

There is a squamous epithelium cyst-like structure with solid areas, formation of squamous eddies, and occasionally atypical cytologic features (Fig. 17.19). The key to the diagnosis is the presence of trichilemmal keratinization. This tumor represents a proliferating trichilemmal cyst, and some may act malignant. Complete excision is recommended.

Malignant Hair Follicle Tumors

Trichilemmal carcinoma is a malignant epithelial tumor with hair follicle differentiation. In order to establish the diagnosis a clear connection with a hair follicle should be found. **Pilomatrix carcinoma** refers to a pilomatricoma in which there are mitoses, some of them atypical, invasion of the surrounding dermis and atypical cytologic features.

17

CHAPTER 18

Sebaceous, Eccrine and Apocrine Tumors

Angela Yen Moore

Section 49. Sebaceous Gland Tumors

Nevus Sebaceous

Clinical

It is a congenital benign but persistent yellow patch or papillomatous plaque with alopecia. At birth, maternal hormonal stimulation may prompt papillomatosis but otherwise nevus sebaceous may be fairly flat until puberty. Nevus sebaceous usually appears on the scalp but also presents on the face, neck, trunk, and extremities. After puberty, hormonal stimulation may induce verrucous growth and increased pigmentation. Syringocystadenoma papilliferum, basal cell carcinoma, tubular apocrine adenoma, or other adnexal tumors may also develop within the lesion most commonly after puberty with increasing incidence with increasing age.

Histology

Epidermal hyperplasia with papillomatosis, numerous apocrine glands, and normal or enlarged sebaceous glands, which may be connected directly to the epidermis unassociated with hair shafts. Before puberty, pilosebaceous units may be poorly developed and small (Fig. 18.1).

Sebaceous Hyperplasia/Rhinophyma

Clinical

Sebaceous hyperplasia refers to flesh-colored to yellow papules with central umbilication that usually occur on the face of older individuals with chronic sun exposure but may appear in adolescence or early adulthood. Confluence on the nose, particularly in individuals susceptible to acne rosacea, may produce enlargement of the nose, or rhinophyma.

Histology

Enlarged sebaceous gland often centered around a large central orifice and associated with solar elastosis. Most cells are sebaceous, with basaloid cells located only at the edges of the sebaceous gland lobules.

Dermatopathology, edited by Ramón L. Sánchez and Sharon S. Raimer. ©2001 Landes Bioscience.

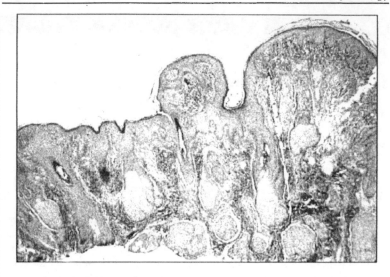

Fig 18.1. **Nevus sebaceous**: There is papillomatosis of the epidermis, and abundant sebaceous glands, some of which open directly into the epidermis.

Fordyce's Condition

Clinical
Ectopic sebaceous glands presenting as pinpoint yellow papules on the vermilion border of the lips or on the oral mucosa.

Histology
Identical to that seen in sebaceous hyperplasia.

Sebaceous Adenoma

Clinical
Solitary, tan-yellow papule usually less than 1 cm in size that appears on the face of chronically sun-exposed middle-aged adults; when associated with Muir-Torre syndrome, often are multiple. Muir-Torre syndrome is also associated with multiple sebaceous epitheliomas, sebaceous carcinomas, basal cell carcinoma, squamous cell carcinoma, multiple keratoacanthomas, and visceral carcinomas. These patients are especially prone to colon carcinoma.

Histology
Sebaceous lobules with more than 50% sebaceous cells and less than 50% basaloid cells. The sebaceous lobular architecture is not disrupted (Fig. 18.2).

Sebaceous Epithelioma
(Basal cell carcinoma with sebaceous differentiation)

18

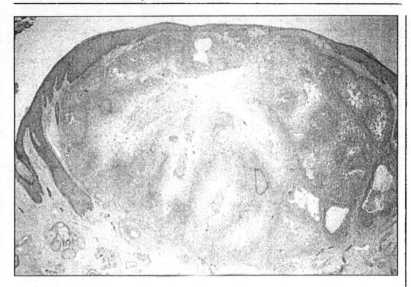

Fig 18.2. **Sebaceous adenoma**: A well delimited tumor nodule made up of approximately equal proportions of clear, sebaceous cells and basaloid cells.

Clinical
Similar in appearance to sebaceous adenoma but may bleed and ulcerate.

Histology
More than 50% of the cells are basaloid and less than 50% are of sebaceous differentiation. The architecture of the sebaceous lobule is lost; instead, lobules of intermixed basaloid cells with focal areas of sebaceous differentiation are seen (Fig. 18.3).

Sebaceous Carcinoma

Clinical
Usually presents as a yellowish tumor on the face or less commonly on the scalp, neck or elsewhere on the body. The lesion is locally invasive and occasionally produces distant metastases. On the eyelids they originate from the meibomian or zeiss glands.

Histology
Poorly defined lobules of basaloid cells and immature sebaceous cells infiltrating the dermis and sometimes associated with pagetoid cells in the epidermis. Less than 10% of the cells are sebaceous cells and areas of keratinization exist so that differentiation from squamous cell carcinoma may be difficult. Severe atypia with hyperchromatic, pleomorphic cells with numerous mitoses as well as a chronic inflammatory infiltrate as a host response are present (Fig 18.4).

18

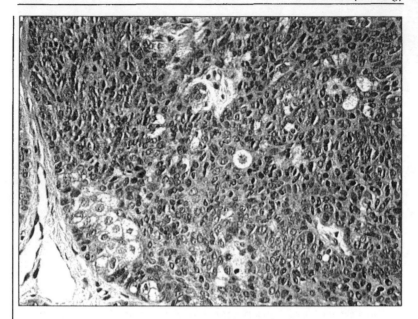

Fig 18.3. **Sebaceous epithelioma**: Tumor nests of basaloid cells, resembling a basal cell carcinoma, admixed with sebaceous cells (BCC with sebaceous differentiation).

Fig 18.4. **Sebaceous carcinoma**: A squamous cell carcinoma with sebaceous differentiation. Pagetoid invasion of the epidermis by sebaceous cells is likely in these tumors.

18

Section 50. Eccrine Gland Tumors

Eccrine Hidrocystoma

Clinical
Solitary flesh-colored to translucent papule which may be slightly bluish- tinged. The most common location is the face. Multiple lesions in cooks have been reported and attributed to repeated high temperature exposure.

Histology
Large unilocular cyst located in the mid-dermis and surrounded by edematous, hemorrhagic stroma. No papillary infolding of the cyst wall is usually present; instead, the cyst wall is lined by two layers of cuboidal cells. Small eccrine glands and ducts may be located below the cyst (Fig. 18.5).

Eccrine Poroma and Porocarcinoma

Clinical
Eccrine poroma is a solitary nodule that most commonly (60%) presents on the sole of the foot or the palms but also appears elsewhere, such as on the head, neck, and trunk. Affected individuals are usually middle-aged, and lesions are usually painless, firm to rubbery, dome-shaped, or rarely pedunculated nodules less than 2-3 cm in diameter. Multiple and diffuse lesions may be associated with hidrotic ectodermal dysplasia. Porocarcinoma usually presents as a solitary nodule or plaque with ulceration on the head or extremities of elderly patients. Metastases often occurs to the skin as well as viscera.

Histology
There is a plate-like dermal nodule with multiple, broad connections to the epidermis. Sharp demarcation is present between normal epidermis and the tumor lobule, which consists of epidermal acanthosis with cuboidal or basaloid cells within the lower epidermis and without a palisading outer layer. Tumor cells may be clear from glycogen accumulation. Small sweat ducts and cystic spaces in the tumor may be present (Fig. 18.6). Tumor masses are embedded in a vascular stroma. Mitoses indicate malignancy and development of porocarcinoma. Porocarcinoma may display pagetoid spread.

Stains
PAS positive, diastase labile

Hidroacanthoma Simplex

Clinical
Resembles macular seborrheic keratosis or Bowen's disease.

18

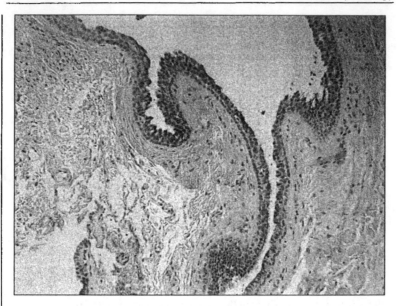

Fig 18.5. **Eccrine hidrocystoma**: A cyst lined by two rows of cells, one of them basaloid and the other cuboidal to cylindrical.

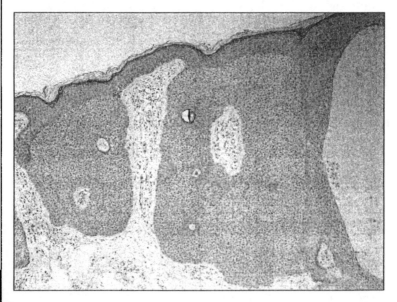

Fig 18.6. **Eccrine poroma**: Tumor made up of basaloid cells, connected to the epidermis. Cysts or duct lumina are commonly seen within the tumor.

18

Histology

Circumscribed islands of cuboidal cells resembling those of eccrine poroma but restricted to an acanthotic epidermis without dermal involvement. Tumor cells contain glycogen.

Stains

PAS positive and diastase labile.

Dermal Duct Tumor

Clinical

Solitary nodule that usually arises on the head, neck, and back of middle-aged and older women.

Histology

Variant of eccrine poroma in which tumor is located completely in the dermis without epidermal connection and without a prominent vascular stroma.

Eccrine Acrospiroma/Clear Cell Hidradenoma

(Nodular Hidradenoma, Solid-Cystic Hidradenoma)

Clinical

Solitary asymptomatic intradermal nodule that varies in color from flesh-colored to red to blue and measures between 0.5 and 2 cm in diameter. No site predilection exists, and involvement of the face, scalp, chest, and abdomen have been reported.

Histology

Nodular dermal tumor with possible epidermal connection. It is composed of two cell types of basaloid cells and sweat duct lumina which vary in size from small ducts to large cystic spaces. Solid portions of the tumor consist of small basaloid and squamoid polyhedral or fusiform cells (Fig. 18.7). Characteristic areas of hyalinized collagen may form mantles around the tumor nodules and extend into the tumor as trabeculae.

Eccrine Spiradenoma

Clinical

Painful, flesh-colored, 1-2 cm deep nodules in children and young adults. Lesions may be solitary, multiple, or linear. No site of predilection exists, although most have been reported on the trunk and extremities.

Histology

Sharply demarcated dermal nodule(s) without epidermal connection with two cell types of basaloid cells arranged in rosettes. Sweat duct lumina may be present (Fig. 18.8).

18

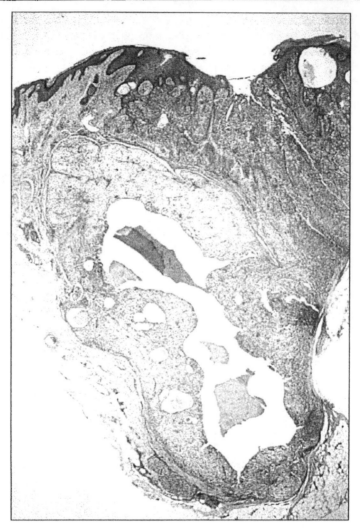

Fig 18.7. **Eccrine acrospiroma**: Also known as clear cell hidradenoma, is commonly cystic, often connected to the epidermis and showing a mixture of cells, clear to basaloid.

Syringoma

Clinical

Flesh-colored to tan pinpoint papules that appear usually on the face and chest, most commonly on the lower eyelids and upper cheeks. Women are far more commonly affected. Eruptive syringomas may be diffuse, linear, or unilateral. Syringomas

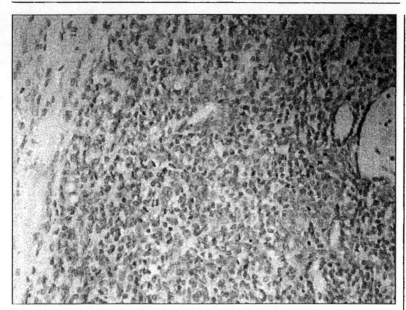

Fig 18.8. **Eccrine spiradenoma**: Nests of tumor cells often in lower dermis or subcutaneous tissue. Two types of cells, larger clear and small basaloid. Duct formation typical.

may occur in association with Down syndrome, Marfan syndrome, and Ehlers Danlos syndrome.

Histology

Tadpole-shaped ductal structures in the dermis without epidermal connection. Surrounding stroma may be fibrotic. May possess cystic spaces containing keratin. Variant may contain clear cells containing glycogen (PAS positive, diastase labile) (Fig. 18.9).

Cylindroma

Clinical

Flesh-colored nodule that appears on the face and scalp either as the solitary type during adulthood or as the multiple type occurring at puberty as exophytic nodules in turban areas ("turban tumors"). Multiple cylindromas can occur with multiple trichoepitheliomas as an autosomal-dominant inherited trait. Cylindromas may also be associated with eccrine spiradenomas and membranous basal cell adenoma of the parotid gland.

18

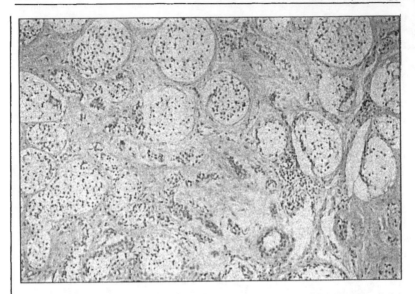

Fig 18.9. **Clear cell syringoma**: Nests of clear cells, glycogen rich, in the superficial dermis. Ducts with tadpole arrangement, characteristic of classic syringoma, are also seen.

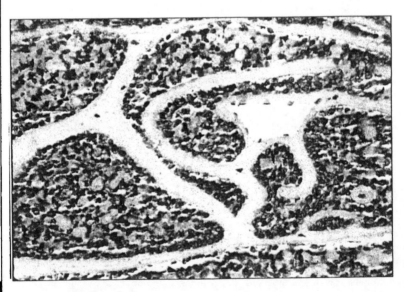

Fig 18.10. **Cylindroma**: Nests of basaloid cells fitting like pieces of a jigsaw puzzle, separated by a thick eosinophilic basal membrane. Small spaces are also seen in the nests.

Histology

Tumor islands in the dermis composed of two cell types of basaloid cells. Fitting together like a jigsaw puzzle, each tumor island is surrounded by a hyalinized cylinder of thickened basement membrane and contains hyalinized droplets and/or sweat duct lumina. Occasionally, cylindromas may extend into the subcutaneous fat but should not show atypia (Fig. 18.10).

Stains

Hyalinized membranes are PAS-positive and diastase-resistant. Tumor cells stain with actin or S-100, or both, consistent with myoepithelial and secretory eccrine gland differentiation.

Chondroid Syringoma (Mixed Tumor of the Skin)

Clinical

Solitary flesh-colored 0.5-3 cm firm nodule most commonly found on the head and neck.

Histology

Tadpole-like ductal structures in the dermis, without epidermal connection, surrounded by prominent mucinous stroma often with chondroid areas and with occasional hyalinized areas (Fig. 18.11).

Stains

Mucinous stroma stains positive with such acid mucopolysaccharide stains as alcian blue at pH 2.5 and 0.5 and toluidine blue at pH 2.5 and 0.5.

Papillary Eccrine Adenoma

Clinical

Solitary nodule ranging in size from 0.5 to 2 cm that is usually found on the extremities and more commonly seen in blacks. This entity is the counterpart of tubular apocrine adenoma.

Histology

Dermal lobules without epidermal connection consisting of glandular spaces lined by two cell layers and papillary projections into the lumina. Outer layer consists of flattened or cuboidal cells. Inner cells may be flattened, cuboidal, or columnar. The cystic lumen contains an amorphous eosinophilic secretory material.

Stains

Ductal structures stain positively with S-100, carcinoembryonic antigen, and epithelial membrane antigen.

18

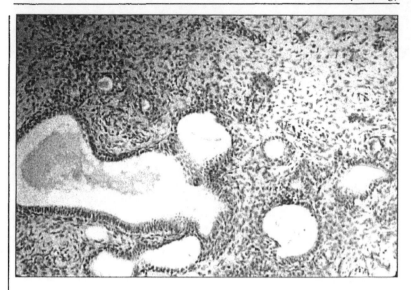

Fig 18.11. **Chondroid syringoma**: Glandular spaces and ducts embedded in a fibrous stroma rich in mucopolysaccharides and occasionally forming cartilage.

Microcystic Adnexal Carcinoma (Sclerosing Sweat Duct Carcinoma, Malignant Syringoma, Syringoid Eccrine Carcinoma, Syringomatous Carcinoma)

Clinical

Solitary papule or indistinct plaque that most commonly arises on the upper lip, nasolabial area, or periorbital region although reports on the axilla, trunk, and scalp also exist. Patients range in age from 18-76 years. Growth rate appears slow although local recurrence is common.

Histology

Deeply infiltrating dermal strands of ductal cells with small lumina and occasional horn cysts that may mimic syringomatous ducts. Perineural invasion is frequent, while atypia with prominent mitoses, hyperchromatism, and pleomorphism may be minimal and necrosis is rare (Fig. 18.12).

Adenoid Cystic Carcinoma

Clinical

Primarily occurring in middle-aged to elderly women, painful or hyperesthetic plaques, often exceeding 3 cm in diameter, affect the scalp, trunk, and extremities most commonly.

18

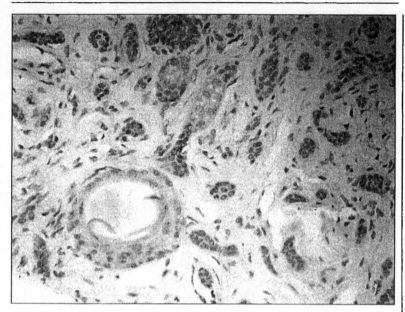

Fig 18.12. **Microcystic adnexal carcinoma**: Bland appearing tumor formed by ducts, epithelial microcysts and cords of cells, infiltrating deeply in the dermis.

Histology
Dermal ducts forming glandular structures, often cribiform, and containing mucin infiltrate the dermis without epidermal connection. Atypia often present with prominent mitoses, hyperchromatism, and pleomorphism.

Mucinous Carcinoma

Clinical
Skin-colored to blue-pink plaques or nodules affecting middle-aged to elderly males (2:1). Most commonly (80%) presenting on the head or face, these solitary nodules affect the eyelids in 50% of cases.

Histology
Infiltrative dermal lobules without epidermal connection consisting of small ductal structures embedded in abundant pools of mucin and separated from each other by fibrous septae. Nuclear atypia often present (Fig. 18.13).

Stains
The mucinous material is composed of nonsulfated epithelial mucin positive for alcian blue at pH 2.5, PAS, mucicarmine, and colloidal iron.

18

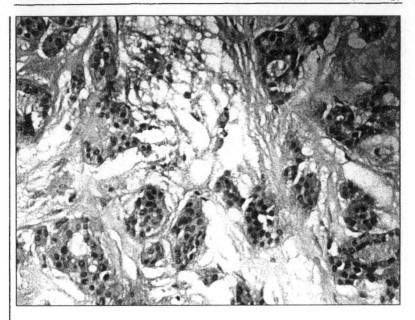

Fig 18.13. **Mucinous carcinoma**: Small nests of epithelial cells and ducts admixed in a clear mucinous stroma. Mucicarmin stain.

Eccrine Carcinoma

Clinical
Rare, eccrine carcinoma represents a less differentiated form of other carcinomas of eccrine derivation and possesses a high rate of metastatic spread.

Histology
Ductal or glandular proliferations infiltrating the dermis without epidermal connection. Atypia with hyperchromatism, pleomorphism, and numerous mitoses is present.

Section 51. Apocrine Gland Tumors

Apocrine Hidrocystoma (Apocrine Cystadenoma)

Clinical
Solitary blue-black nodule on the face, head, neck, and upper trunk that ranges in size from several millimeters to several centimeters and usually affecting adults in their sixth decade.

18

Histology

Single or multiple cystic cavity in the upper dermis that is lined by an outer single myoepithelial cell layer and multiple cuboidal to columnar cells exhibiting decapitation secretion.

Apocrine Nevus

Clinical

Localized lesion in the scalp, axilla, presternal, or inguinal area that may also be seen in association with nevus sebaceous or syringocystadenoma papilliferum.

Histology

Increased size or number of mature apocrine glands, sometimes associated with epidermal basaloid hyperplasia.

Hidradenoma Papilliferum

Clinical

Solitary, mobile nodule in the labial or perianal region in women which may exhibit superficial erosion and moisture. Occurrence at the nipple, eyelid, and external ear canal have been reported. Classically, this condition affects white women over the age of 30.

Histology

Dermal circumscribed tumor without epidermal connection and with minimal inflammation around the tumor. Glandular tumor containing cystic spaces with papillary projections lined by cells showing decapitation secretion (Fig. 18.14).

Syringocystadenoma Papilliferum

Clinical

Papillomatous plaque that arises most commonly on the scalp at the site of a preexisting nevus sebaceous but may appear by itself on the head or neck or elsewhere.

Histology

Tumor connected to papillomatous epidermis and containing a cystic space which opens to the surface. Papillary projections into the cystic space line the luminal surface, with luminal cells being squamous at the upper portion and two-cell thick at the lower portion. The lower portion contains luminal columnar cells with decapitation secretion and peripheral cuboidal cells. Plasma cells in the surrounding stroma are present (Fig. 18.15).

18

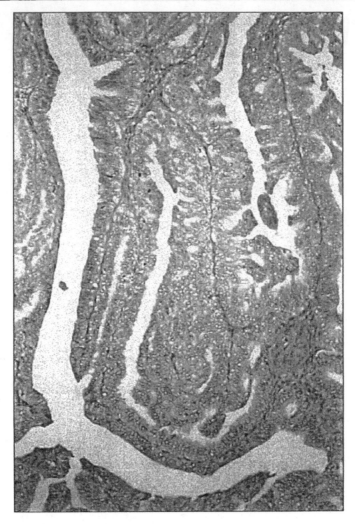

Fig 18.14. **Hidradenoma papilliferum**: Well delimited tumor nodule made up of papillary fronds lined by cylindrical epithelial cells.

Erosive Adenomatosis of the Nipple (Florid Papillomatosis of the Nipple Ducts, Nipple Adenoma)

Clinical

Nipple shows nodular enlargement with superficial erosions and blood-stained or serous discharge.

18

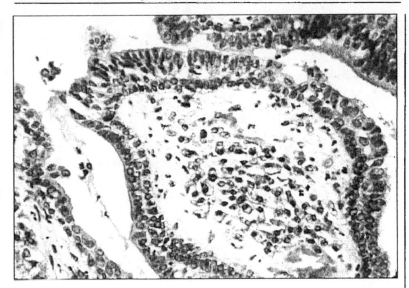

Fig 18.15. **Syringocystadenoma papilliferum**: Papillary formations with wide fibrovascular stalk in which plasma cells are typically seen. Papillae lined by two rows of cells.

Histology

Dermal tumor with frequent epidermal connection and consisting of glandular spaces with papillary projections into the lumina and lined by cells showing decapitation secretion. Dermal infiltrate with mononuclear cells or plasma cells may surround the tumor. Can resemble syringocystadenoma papilliferum histologically.

Stains

Positive for antikeratin, carcinoembryonic antigen, and epithelial membrane antigen stains.

Tubular Apocrine Adenoma

Clinical

Solitary, flesh-colored, dome-shaped or pedunculated nodules most commonly present at the perianal and axillary skin. May occur within a pre-existing nevus sebaceous of the scalp.

Histology

Symmetrical, well-circumscribed, dermal tumor without epidermal connection and with variable dilated and branching tubules within the dermis. These nests of glandular spaces contain papillary projections into the lumina that are lined by columnar luminal cells showing decapitation secretion and cuboidal peripheral cells. Nests are separated by thin strands of fibrous septae.

18

Differential Diagnosis
Metastatic adenocarcinoma with apocrine features is asymmetrical, contains mitoses, and exhibits cellular necrosis.

Apocrine Fibroadenoma

Clinical
Flesh-colored nodule at the vulvar or perianal skin that is often multiple.

Histology
Resembles intracanalicular fibroadenomas of the breast but lined by cells exhibiting decapitation secretion.

Apocrine Carcinoma

Clinical
Red to purple cyst often with overlying ulceration. It more commonly appears in males and in such apocrine areas as the axilla and anogenital skin, although such apocrine gland areas as the ear canal and eyelids may also be affected.

Histology
Cyst formation in hemorrhagic areas may be present, with focal decapitation secretion and cribiform patterns of luminal bridging at some glandular areas. Poorly differentiated areas display solid sheets or nests of cells with abundant eosinophilic cytoplasm but may otherwise be indistinguishable from metastatic adenocarcinoma. Numerous mitoses, atypical mitoses, and necrosis as well as stromal fibrosis and hyalinization are often present. Normal apocrine glands may often be found nearby.

Stains
PAS-positive diastase-resistant and Perl's stain (iron) positive granules.

CHAPTER 16

Soft Tissue Tumors

Ramón L. Sánchez and Sharon S. Raimer

Section 52. Smooth Muscle Tumors/Striated Muscle

Smooth muscle cells and bundles are seen in the dermis as part of the arrector pilaris muscle. Bundles of smooth muscle cells are also seen randomly distributed in the dermis, in the area of the areola, as well as in the scrotum (dartos). Smooth muscle tumors in the skin originate from these normal anatomical locations and also from the smooth muscle cells from the walls of arteries and veins.

Leiomyoma

Clinical
Small nodules, can be solitary or multiple. They may appear anywhere in the body and are painful.

Histology
A well-circumscribed collection of smooth muscle cells. Usually larger and rounder than arrector pilaris muscle. On longitudinal section the nuclei appear in the center of the sarcolemma. Smooth muscle bundles , when cut across, show characteristic vacuolation, which allows recognition of the tissue. Angioleiomyomas are leiomyomas originating in the smooth muscle cells in the walls of blood vessels. They often recapitulate the architecture of the blood vessel, being completely round, well-circumscribed and presenting with a small central lumen. Angioleiomyomas are often multiple (Fig. 19.1).

Stains
Smooth muscle cells stain red with trichrome. With immunoperoxidase stains they are positive for smooth muscle actin and vimentin.

Differential Diagnosis
Leiomyomas should be differentiated from smooth muscle hamartomas. The latter show smooth muscle bundles that are randomly distributed throughout the dermis without clear demarcation.

Myofibromatosis, Infantile
This entity appears as congenital tumors or develops in early childhood. It presents as small nodules or firm plaques, which can be single or multiple, and are usually confined to the skin or subcutaneous tissue. Occasionally myofibromatosis also

---

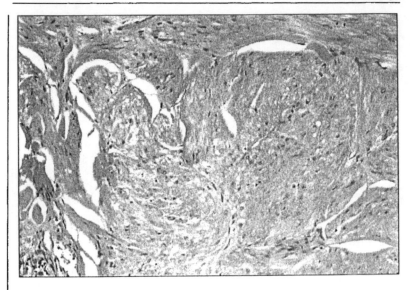

Fig 19.1 . **Leiomyoma**: A tumor nodule formed by spindle cells arranged in bundles, some of which are cut across. Cross sectioning show vacuoles in their cytoplasm.

involves other organs, such as the heart, lung or liver, and in which case the prognosis is poor. Solitary myofibromatosis has also been described in adults arising de-novo.

Histology

Collections of myofibroblasts are seen in the dermis. The histologic appearance resembles leiomyomas, although the cells are less defined and mature. Myofibroblasts stain weakly with smooth muscle actin.

Leiomyosarcoma

Clinical

Tumor nodules that may arise in any location, often with rapid growth. Primary skin tumors are seen mostly in adults on the limbs, especially the legs, and are more common in males. Superficial tumors have better prognosis than those that involve deeper tissues.

Histology

A spindle cell tumor, which often involves the complete thickness of the dermis up to the epidermis. Although there is not a characteristic architectural pattern, areas of storiform arrangement can be seen. The individual cells display an elongated vacuolated nucleus and spindle cytoplasm. Mitotic figures, usually regular, are frequently seen. Occasional tumor cells may present pleomorphism and atypia, although never to the degree seen in atypical fibroxanthoma. The differential diagnosis

19

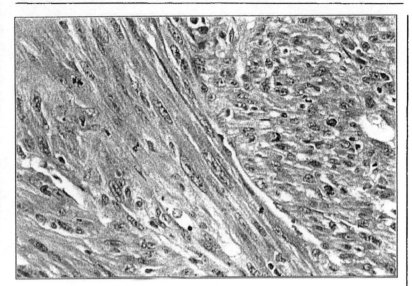

Fig 19.2. **Leiomyosarcoma**: The spindle, smooth muscle cells show large, atypical nuclei with prominent nucleoli. Mitotic figures are frequently seen.

includes spindle cell squamous cell carcinoma, spindle cell melanoma, as well as atypical fibroxanthoma. Leiomyosarcomas characteristically stain positive with smooth muscle actin (Fig. 19.2).

Rhabdomyosarcoma

Clinical

There are two main settings, one in small children, characterized by nodules that are seen more often on the head and neck or in the genito-urinary region. The other setting is in adolescents and adults, characterized by large soft-tissue tumors that are deep and involve the skin secondarily.

Histology

The tumors seen in children are undifferentiated tumors that can be included in the general category of small blue cell tumors. Many of them are embryonary rhabdomyosarcomas, and are characterized by small cells with hyperchromatic nuclei and scanty cytoplasm (Fig. 19.3). Occasional strapp cells or tadpole cells, if seen, are diagnostic by showing cross striations. The latter can be demonstrated easily after staining with PTAH (phosphotungstic acid hematoxillin). Some rhabdomyosarcomas in children, although quite undifferentiated, can be classified as alveolar rhabdomyosarcomas. The rhabdomyosarcomas seen in adults are either a classic alveolar or pleomorphic rhabdomyosarcomas. The latter tumors show marked atypia, bizarre cells, necrosis, and abundant mitoses.

19

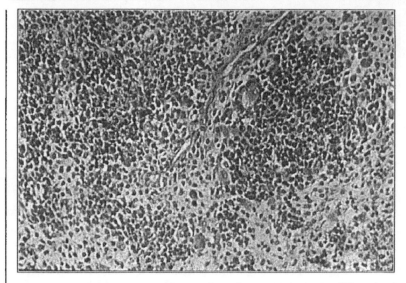

Fig 19.3. **Embryonal rhabdomyosarcoma**: Blue cell tumor made up of undifferentiated cells with scanty cytoplasm and homogeneous stroma. Strap cells are characteristic.

Stains
Myoglobin and actin stains may be positive in these tumors.

Section 53. Fatty Abnormalities and Tumors

Lipoma and Angiolipoma

Clinical
· They are the most common soft tissue tumors. They can be seen anywhere on the body but are most common on the upper extremities and trunk, where they appear as soft, depressible, deep-seated nodules that should be differentiated from epidermal cysts. They may be somewhat translucent upon illumination. Usually they are asymptomatic. If they are painful, the possibility of angiolipoma should be considered. Although often solitary, it is not uncommon to find two or three in the same patient. Lipomatosis dolorosa, or Dercum's Disease, refers to multiple diffuse lipomas, which are symptomatic.

Histology
Lipomas may be surrounded by a thin capsule but usually appear as relatively well-circumscribed collections of mature fat. Histologically a lipoma can not be clearly separated from normal subcutaneous fat. If there are many capillaries, some of them with fibrinoid thrombi, the diagnosis of angiolipoma is usually made (Fig. 19.4). Although they are usually subcutaneous, lipomas occasionally involve part of

19

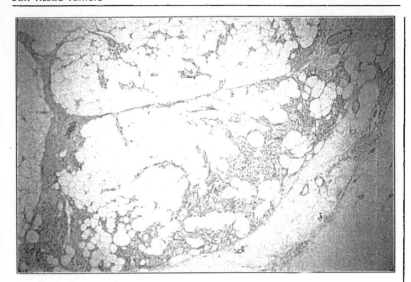

Fig 19.4. **Angiolipoma**: Mature adipose tumor with a prominent capillary network. Some of the capillaries show fibrin thrombi in their lumen.

the dermis. In that situation, they should be differentiated from lipofibromas, nevus lipomatosus superficialis, and from Goltz syndrome (focal dermal hypoplasia).

Subtypes

Spindle cell lipomas show a considerable number of spindle cells resembling fibroblasts without a clear architectural arrangement admixed among the adipocytes. **Pleomorphic lipomas** show atypical cells, including multi-nucleated giant cells, which make the tumors suspicious for liposarcoma. The differentiation is established by the absence of lipoblasts, mitosis and necrosis. **Hibernoma** refers to a tumor made of fetal adipocytes, which show a central nucleus and multiple small vacuoles filling up the cytoplasm of the cell. Fetal fat is normally seen after birth in the adrenal glands. Hibernomas usually occur in the trunk and extremities (Fig. 19.5). **Lipoblastoma** refers to the occasional finding usually in children, of collections of lipoblasts. The lipoblasts may be admixed with normal mature adipocytes, and they are characterized by a central nucleus and multivesicular cytoplasm. Lipoblastomas are often incidental findings when studying soft tissue from surgical procedures, such as herniorrhaphy specimen.

Liposarcomas

Clinical

Usually larger than lipomas, with a diameter of at least five centimeters, they may present anywhere on the trunk or extremities. Occasionally a history of rapid growth can be elicited.

19

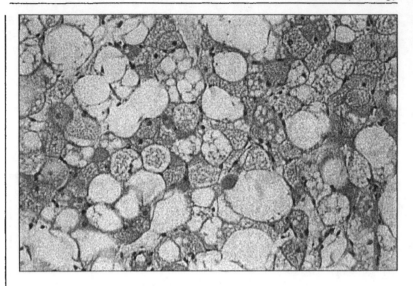

Fig 19.5. **Hibernoma**: A tumor with a large proportion of brown fat cells, characterized by microvesicular, foamy cytoplasm and smaller size than mature adipocytes.

Histology

Liposarcomas can be roughly divided into low and high grades. Well-differentiated liposarcoma resembles a lipoma with an occasional lipoblast and an increased vascular pattern. There is usually a history of large recurrent lipoma. Myxoid liposarcomas are characterized by spindle cells in a mucinous stroma admixed with normal adipocytes, and by a prominent vascular pattern. The blood vessels branch, producing the so-called **chicken feet** appearance (Fig. 19.6). A requirement for diagnosis of liposarcoma is the presence of lipoblasts. They can be seen as multi-vacuolated cells, or as cells with a single cytoplasmic vacuole and large hypocromatic oval nucleus. The high grade liposarcomas, or pleomorphic liposarcomas, are characterized by large atypical cells with pleomorphic nuclei, abundant mitosis, many of them atypical, surrounded by occasional adipocytes or lipoblasts. These tumors are not well-differentiated from other pleomorphic sarcomas and they behave, and are also treated, in a similar fashion.

Section 54. Mesenchymal Tumors

Mesenchymal tumors of the skin constitute a heterogeneous group of disorders that include hamartomas and other congenital conditions, acquired benign tumors, and some premalignant or malignant neoplasms. The histologic constituents of all of these entities are also heterogeneous and include the fibroblasts and the dermal dendrocytes, as well as other mesenchymal cells of the dermis. Different degrees of collagenization, different amounts of ground substance, myofibroblasts, as well as

19

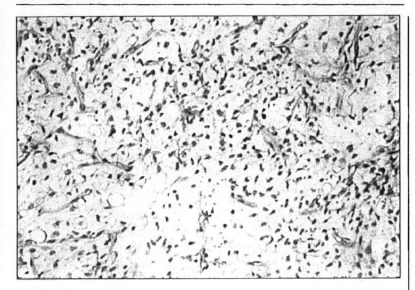

Fig 19.6. **Myxoid liposarcoma**: A fatty tumor with a myxoid background, a characteristic vascular pattern (chicken feet), and the presence of lipoblasts.

fat may be present. Some tumors such as dermatofibrosarcoma protuberans are made of spindle cells whose nature is not completely elucidated.

Dermatofibroma

Clinical

Very common tumors, which may be single or multiple. Mainly seen on extremities, particularly thighs and legs, although they can be seen anywhere. They present as small reddish nodules that grow to a size generally of less than one centimeter. They become darker in color with time, while flattening. By pinching the skin around the dermatofibromas, the dimple sign can be elicited.

Histology

Spindle infiltrate involving most of the thickness of the dermis. Cutaneous adnexa are usually displaced at the periphery of the tumor, although occasionally they may be surrounded by the infiltrate. The constituents of dermatofibroma are variable and are made of a mixture of spindle cells, histiocytic-like cells with more abundant cytoplasm, multi-nucleated giant cells, some of which are Touton giant cells, and blood vessels, as well as collagen bundles. Depending on the predominance of each of those components, dermatofibromas have been classified into cellular and fibrous dermatofibromas (Fig. 19.7). The former show a variety of cell constituents with little collagen, while the latter exhibit spindle cells and abundant collagen. The relative predominance of each of these cellular components is probably a function of the age of the lesion, with young lesions being more cellular, while older ones are

19

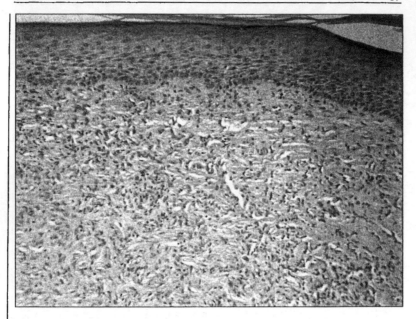

Fig 19.7. **Dermatofibroma**: The epidermis is acanthotic and often hyperpigmented. The dermis shows a tumor with spindle cells, storiform pattern, and inflammation.

mainly fibrous. Likewise, it is not uncommon to see in a dermatofibroma a fibrous collagenized center with a more cellular periphery.

The epidermis in dermatofibromas is characteristically hyperplastic and shows hyperpigmentation of the basal layer. The acanthosis of the epidermis can produce clusters of basaloid cells that resemble superficial basal cell carcinoma. The epidermal hyperplasia may be triggered by the dermal infiltrate reaching into the basal layer of the epidermis. Two important histologic features help in the recognition of this entity. One is the presence of keloidal collagen, hyalinized bundles of collagen seen among the infiltrate. The other feature is the infiltrative nature of the tumor cells among the collagen bundles of the dermis at the periphery of the lesion.

Special Stains

Factor XIII-a will stain most of the cells of the dermatofibroma. The same stain in older lesions may show a negative center with a positive stain in the cells of the periphery.

Dermatofibroma Variants

Hemosiderotic dermatofibroma refers to those lesions with abundant blood vessels and hemorrhage which exhibit heavy deposits of hemosiderin as a result of previous bleeding. The lesions on low magnification may resemble a melanocytic pigmented lesion. This lesion was previously referred to as a sclerosing hemangioma

(Fig. 19.8). A related entity is **aneurysmal dermatofibroma**, which shows large dilated blood spaces, particularly in the deeper areas. These lesions clinically are sometimes diagnosed as hemangiomas or epidermoid cysts. **Cholesterolotic dermatofibroma** refers to a dermatofibroma which partially or completely has experienced transformation of the infiltrate into foamy macrophages resembling a xanthoma. The cholesterolotic change is probably related to high blood lipids, and should be considered a warning sign for such. **Dermatofibroma with monster cells**, or atypical dermatofibroma, refers to such lesions with large, atypical giant cells, some of which may be multi-nucleated. The so-called **sclerotic fibroma of the skin** is probably the end result of a dermatofibroma, and it is characterized by well-defined collagenized and hypocellular nodules in the dermis. Some of these lesions have been localized on the scalp. **Dermatomyofibroma** refers to a lesion that is seen often on the shoulder area, particularly in younger women, characterized by bundles of spindle cells without a storiform pattern and parallel to the surface resembling a fibrous dermatofibroma.

Most of the cells are probably myofibroblasts. **Subcutaneous benign fibrohistiocytoma** represents a dermatofibroma in the subcutaneous tissue. It is usually composed of mixture of cells similar to those seen in cellular dermatofibroma. In these cases, as in older dermatofibromas, mitoses are occasionally seen, which do not imply any change in prognosis.

Fibrous Hamartoma of Infancy

Clinical

Fibrous hamartoma of infancy is usually a solitary tumor seen in young children, often on extremities. It may reach a size of five or six centimeters in diameter.

Histology

The dermis, as well as the subcutaneous tissue, is partially replaced by a mixture of three tissue components: mature fat, bundles of mature fibrous tissue (collagenized), and groups of cells that resemble embryonal collagen with cells that display spindle nuclei and scanty cytoplasm embedded in a mucinous stroma. A pentachrome stain shows the different components clearly (Fig. 19.9).

Fibrous Papule of the Nose/Angiofibromas

Clinical

Fibrous papules are very common, usually solitary, about two to three millimeters in average diameter, particularly seen on the nose, near the tip or on the alae. Fibrous papules, however, are also seen elsewhere on the face. Angiofibromas, synonymous for fibrous papules, can be multiple, particularly distributed bilaterally near the nasal labial fold in tuberous sclerosis. Angiofibromas may also be seen in the latter entity and as a related finding on the fingers and toes, particularly near the nail on the lateral fold.

19

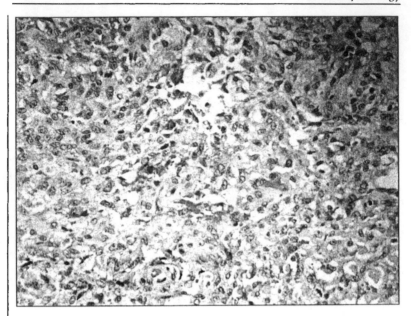

Fig 19.8. **Hemosiderotic dermatofibroma**: Cellular dermatofibroma with abundant blood vessels, extravasated red blood cells, macrophages and hemosiderin deposition.

Fig 19.9. **Fibrous hamartoma of infancy**: The three histologic components include collagenized fibrous tissue, adipose tissue and mesenchymal, immature tissue.

19

Histology

Both fibrous papules and angiofibromas are characterized by a papular architecture. The dermis shows elongation of the rete ridges with a decrease of hair follicles and sebaceous glands and abundant dilated blood vessels admixed with larger stellate fibroblasts. The collagen of the dermis may be organized perpendicular to the surface. The angiofibromas in tuberous sclerosis show an absence of elastic fibers, as seen with special stains (VVG or Movat) (Fig. 19.10).

Acquired Digital Fibrokeratoma

Clinical

They are seen in adults, as opposed to recurrent infantile digital fibroma, usually located on the fingers or elsewhere on the hand. They can also be seen on the feet.

Histology

There is hyperkeratosis and acanthosis with a collagenized fibrous dermis which may contain an abundance of blood vessels. Characteristically the collagen bundles in the upper dermis appear to be perpendicular to the surface.

Giant Cell Tumor of the Tendon Sheath

Clinical

They appear as nodules, usually located in the dorsal aspect of the hand, associated with an extensor tendon, either at the level of the phalangeal metacarpal bones, or even as proximal as the wrist. Upon gross examination, the cut surface appears yellowish.

Histology

These are mixed tumors made of histiocyte/macrophagic cells displaying round to oval nuclei and abundant cytoplasm. Foamy histiocytes are responsible for the yellowish color of the tumor. Areas of hemorrhage as well as hemosiderin deposition, and a mixture of inflammatory cells including lymphocyte and plasma cells are present (Fig. 19.11). The histologic findings are very similar to those seen in pigmented villonodular synovitis, although with a different location and without the papillary projection seen in this entity. Fibroma of the tendon may be a related entity, although it is most commonly seen on the palmar aspect of the hand, often associated with the thumb. Histologically fibroma of the tendon is less cellular and more collagenized than the giant cell tumor of the tendon sheath, with absence of giant cells or foamy cells and with collagenized, nodular areas.

Other Mesenchymal Tumors

Elastofibroma dorsa is seen mainly in elderly men in the scapular regions of the back. Histologically it is characterized by a proliferation of elastic fibers throughout the dermis that appear curly and slightly basophilic.

Giant cell epulis occurs mainly on the buccal mucosa, and is characterized histologically by a collection of multinucleated giant cells resembling osteoclasts. **Nodular fasciitis** is a soft tissue proliferation considered to be reactive rather than neoplastic,

19

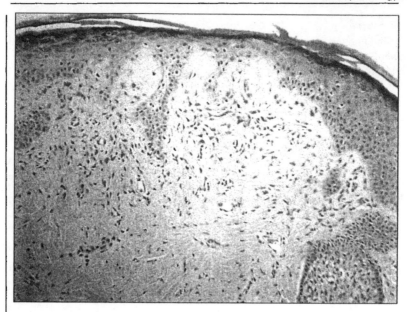

Fig 19.10. **Fibrous papule**: The dermis shows an increased number of capillaries and stellate fibroblasts (angiofibroma).

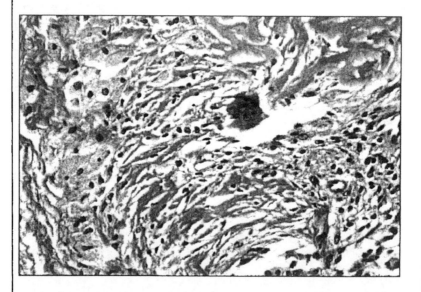

Fig 19.11. **Giant cell tumor of tendon sheath**: Osteoclast-like giant cell, foamy cells (at left), lymphocytes and macrophages. Hemosiderin deposits are also a common finding.

19

seen in young patients, particularly on extensor surfaces near the shoulders, arms, and thighs. There is usually a history of very rapid growth. Histologically nodular fasciitis can be seen connected to the fasciae or in the subcutaneous tissue without any deep connections. It is characterized by a proliferation of spindle cells that appear stellate by displaying cytoplasmic proliferations in a mucinous stroma. There is no architectural pattern, and an infiltrate of lymphocytes is commonly admixed among the tumor cells (Fig. 19.12). Some of the cells of the infiltrate display atypical nuclear features, hence the synonym of pseudosarcomatous fasciitis. The tumor cells are myofibroblasts and stain positive with smooth muscle actin.

Fibromatoses

Clinical

They have been subdivided into six different categories. Those occurring on the palms and soles are usually characterized by round, nodular tumors, which can be clinically very well delineated and mistaken for cysts. Long-standing palmar fibromatosis produce contractions of the digits termed Dupuytren's contractions. Fibromatosis involving the penis is known as Peyronies disease. Abdominal fibromatosis may be seen as a single occurrence or as a part of Gardner's Syndrome. Desmoid tumors refer to fibromatosis of the abdominal wall, often seen during the course of pregnancy. Fibromatosis coli is seen on the neck of children, and it may be congenital.

Histology

Fibromatoses show bundles of mature fibrous tissue with ill-defined borders and an infiltrating pattern at the periphery. Individual tumor cells show spindle wavy nuclei and an abundant collagenized cytoplasm. They show characteristics of collagen and fibrous tissue, and they are negative for CD-34 immunoperoxidase stain. Fibromatoses are considered to be pseudosarcomatous proliferations with a high rate of recurrence due, in part, to the difficulties of complete excision.

Dermatofibrosarcoma Protuberans

Clinical

The classical dermatofibrosarcoma protuberans (DFSP) appears as an elevated, hard multi-nodular tumor, usually seen on the trunk or extremities in middle-aged to elderly patients. The tumor grows relatively quickly, and by the time it is excised, it may have reached five centimeters or more in maximum diameter. A less common variety of DFSP is the atrophic or plaque-like type, which is characterized by an atrophic plaque that may, on occasion, show a focal nodular-like component.

Histology

DFSP is a monocellular tumor characterized by cells with a spindle nucleus and spindle cytoplasm, which is somewhat fibrillar. The cellular infiltrate involves almost the full thickness of the dermis, and it often penetrates into the subcutaneous tissue, usually in bundles of tumor cells. There is typically a storiform or cartwheel archi-

19

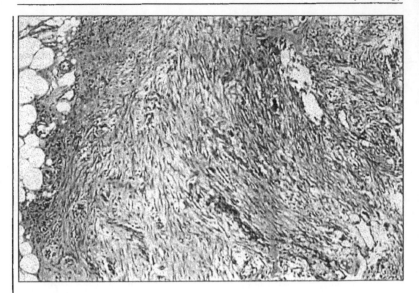

Fig 19.12. **Nodular fasciitis**: Tissue culture-like growth of spindle cells (myofibroblasts), with a myxoid background and ill-defined borders.

tectural arrangement of the tumor cells. The tumor infiltrates between cutaneous adnexa, which are often preserved. Mitoses, although mostly typical, are frequently seen. If there is an inflammatory infiltrate, it is never too prominent (Fig. 19.13). Bednar's Tumor or pigmented DFSP refers to those tumors that have prominent deposits of melanin among the tumor cells. Some DFSPs have shown areas of malignant transformation, which usually display the histologic features of fibrosarcoma; therefore a combination of fibrosarcoma and DFSP is possible.

Special Stains
DFSP characteristically stain positive for CD-34, and this stain is almost diagnostic of the entity. DFSPs are negative for factor XIII-a.

Additional Comments
DFSPs are borderline malignant tumors that require large and deep excisions to avoid recurrence. Grossly, the cut section shows a homogeneous whitish appearance with a well-delineated tumor margin. The delineation seen grossly, however, is misleading since the histologic examination shows infiltration of the surrounding soft tissue by tumor bands. In the past DFSPs have been considered to be related to dermatofibromas. However, DFSPs should be completely separated from dermatofibromas since histologically they show a monomorphic cellular infiltrate, as opposed to the multiple types of cells seen in dermatofibroma. The difference in the nature of the tumors is quite clearly recognized with immunoperoxidase stains, as discussed above.

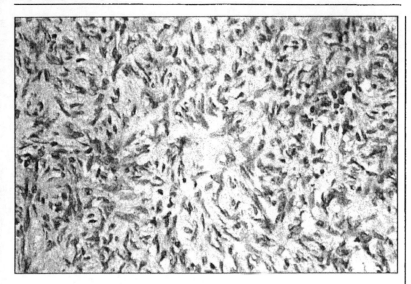

Fig 19.13. **Dermatofibrosarcoma protuberans**: The tumor cells show spindle nuclei, ill defined cytoplasm, homogeneous cytologic features and storiform pattern.

Atypical Fibroxanthoma

Clinical

These tumors are classically seen in older patients on sun exposed areas, particularly the face and forehead. Clinically they are characterized by ill-defined nodules in a setting of solar damage. Younger patients may present these tumors mainly on the thighs or proximal legs.

Histology

Atypical fibroxanthoma is characterized by a proliferation of mesenchymal cells seen in the dermis without any connection to the epidermis. The morphology of the cells is variable, and it ranges from fibroblast-like cells with stellate cytoplasmic proliferations to very atypical mesenchymal cells with hyperchromatic bizarre nuclei. Giant cells are also seen. Mitoses are frequently seen, many of them atypical. In areas the tumor shows a storiform pattern, and hence it resembles histologically malignant fibrous histiocytoma (Fig. 19.14.). The latter, however, are soft tissue sarcomas, which originate in deep soft tissue and never originate in the skin, although the skin can be secondarily involved.

Special Stains

Immunoperoxidase stain for vimentin is positive in the tumor cells. Stains are negative for cytokeratin, S-100, and smooth muscle actin.

Note: Atypical fibroxanthomas are not aggressive tumors and seldom mestastasized, although they recur locally. They are treated by local excision.

19

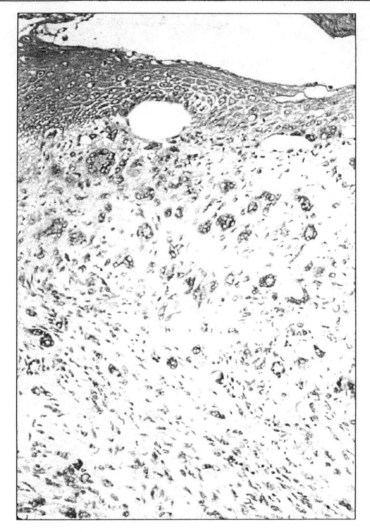

Fig 19.14. Atypical fibroxanthoma: Multitude of atypical, bizarre giant cells, spindle cells with storiform pattern, and atypical mitosis.

Fibrosarcomas

Clinical

Fibrosarcomas are seen occasionally as soft tissue tumor masses, both in adults and in very young children. In the latter setting they may be congenital.

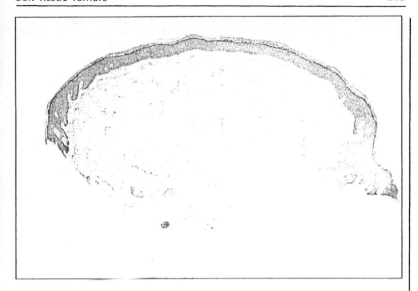

Fig 19.15. **Cutaneous myxoma**: Circumscribed dermal tumor characterized by mucinous stroma, increased amount of capillaries and stellate fibroblasts.

Histology

Histologically they are characterized by spindle tumor cells, which are quite homogeneous without marked pleomorphism. Tumor cells are arranged in small bundles that run in an angle and have been compared to **herringbone**.

Cutaneous Myxomas

Clinical

Small papular tumors, usually less than one centimeter in diameter, which can be hyperpigmented.

Histology

The dermis is replaced by a mucinous material in which stellate-like fibroblasts are seen. Blood vessels may criss-cross the tumor (Fig. 19.15).

Special Stains

The mucinous or mixoid stroma is positive for acid mucopolysaccharides, as seen with alcin blue and colloidal iron stains. Some of the stellate cells stain positive with factor XIII-a with immunoperoxidase staining.

Comments

Cutaneous myxomas should be differentiated from focal mucinosis. Multiple cutaneous myxomas may be part of the Lamb syndrome or Carney's complex. These patients, in addition to cutaneous myxomas, may have cardiac myxomas, and endocrine abnormalities.

19

CHAPTER 20

Pigmentary Disturbances and Melanocytic Tumors

Ramón L. Sánchez and Sharon S. Raimer

Section 55. Pigmentary Disturbances, Nontumoral

Melanin is produced by melanocytes residing in the basal layer of the epidermis. The number of melanocytes has been estimated to be an average of one per ten keratinocytes with some variations throughout the body. The number of melanocytes is fairly constant regardless of race. The differences in color of the skin of different people results from the number, maturity and size of the melanosomes produced by the melanocytes.

The disorders of pigmentation are related to a decreased or increased melanin production or storage. Deposits of other pigmentary substances usually not present in the skin, either endogenous or exogenous, will also result in pigmentation.

Hypomelanosis

Vitiligo/Piebaldism

Clinical
Complete absence of pigment (melanin) in an area of the skin. Vitiligo is an acquired condition thought to be related to destruction of melanocytes by an autoimmune reaction. Piedbaldism is the congenital absence of melanocytes/melanin in a localized area, usually on the forehead and scalp, and in the hair (white forelock).

Histology
Complete absence of pigment and melanocytes throughout the epidermis. A biopsy from the border of vitiligo may show some lymphocytes involving the basal layer.

Stains
S-100, vimentin, Fontana and Dopa reaction will all be negative (absence of melanocytes/melanin)

Differential Diagnosis
Postinflammatory hypopigmentation may show complete absence of melanocytes/melanin.

Dermatopathology, edited by Ramón L. Sánchez and Sharon S. Raimer. ©2001 Landes Bioscience.

Postinflammatory Hypopigmentation

Clinical

It is often associated with scarring, either after trauma or surgery or after an inflammatory dermatosis that results in a scar (i.e., discoid lupus). Burns and cryotherapy may result also in hypopigmentation. Inflammatory dermatosis may cause temporary hypopigmentation.

Histology

Similar to vitiligo. A scar may be present in the dermis. The margin of the hypopigmented macule may show pigment incontinence.

Other Disorders of Hypopigmentation

Hypomelanosis of Ito (Incontinentia Pigmenti Achromians)

Bands and swirls of hypopigmentation generally following the lines of Blaschko. Other congenital abnormalities are common. Histologically decreased number of melanocytes and melanin.

Albinism

Autosomal recessive, it is either tyrosinase positive or negative. Histology shows clear cells in the basal layer that fail to stain with Fontana. Dopa reaction is positive in tyrosinase positive albinism.

Increased Epidermal Pigment

Epidermal Melanosis/Freckles

Clinical

Freckles are small hyperpigmented macules usually on sun exposed areas. Their color increases with sun exposure and there is a congenital predisposition. Occasionally areas of randomly distributed hyperpigmentation are referred to as epidermal melanosis.

Histology

Increased pigmentation of the basal layer without an apparent increase in the number of melanocytes.

Labial Melanotic Macule

Clinical

Darkly hyperpigmented macule usually on the lower lip.

Histology

Hyperpigmentation of the basal layer, often with pigment incontinence (Fig. 20.1).

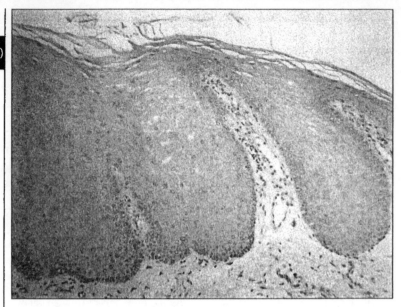

Fig 20.1. **Labial melanotic macule**: Acanthotic squamous mucosa with hyperpigmentation of the basal layer without nest formation or hyperplasia of melanocytes.

Cafe-au-lait Spots

Clinical
Hyperpigmented macules often more than 1.5 cm. in diameter. Patients with neurofibromatosis have at least six cafe-au-lait spots, present since early childhood.

Histology
Increased melanin content in the melanocytes and keratinocytes. Giant melanin granules can be seen in melanocytes.

Differential Diagnosis
Cafe-au-lait spots have clinically smooth borders (coast of California) in contrast to the irregular borders of the melanotic macules in Albright's syndrome (coast of Maine). Giant melanin granules can be seen in both conditions, as well as in melanocytic nevi, lentigines, and even normal skin.

Other Hyperpigmented Disorders

Peutz-Jegher Syndrome
Hyperpigmented macules resembling lentigines in perioral region, lips on oral mucosa, and finger tips. Hamartomatous polyps in intestine. Autosomal dominant.

Melasma

Hyperpigmentation of face, cheeks and upper lip may be associated with pregnancy or oral estrogen. Accentuated by sun exposure. Histologically, hyperpigmentation of basal layer.

Drug Pigmentation

Chlorpromazine

Slate-gray discoloration in exposed areas of the skin. Bulbar conjunctiva may be pigmented. Histologically, brown pigment seen throughout the dermis, in macrophages. The pigment stains positive with Fontana-Masson and bleaches with hydrogen peroxide. Internal organs also show pigment deposition after prolonged therapy.

Minocycline

Diffuse brownish to bluish discoloration accentuated in sun-exposed areas. Acne scars may show blue-black pigmentation. Pigment in acne scars is seen in macrophages and stains like hemosiderin (Perl's stain). The bluish hue on normal skin stains both with Perl's and with Fontana-Masson for melanin.

Hemosiderin Deposits

Stasis Dermatitis

Clinical

Erythematous to purpuric, dermatitis generally associated with edema of the ankles and legs, usually bilateral.

Histology

Changes of chronic dermatitis with increased vascularity in the upper dermis. The capillaries display a lobular arrangement. Hemosiderin granules are seen throughout the dermis, both as free deposits or in macrophages.

Differential Diagnosis

The vascular proliferation in stasis dermatitis is not as prominent as in acroangiodermatitis. The vessels in early Kaposi's sarcoma show an infiltrative tendency and involve also the deep and periadnexal dermis. Both of these entities also show hemosiderin deposits throughout the dermis.

Erythema Ab Igne

Clinical

Reticular erythema, hyperpigmented in chronic cases, following prolonged exposure to heating pads, or other heating devices.

Histology

Increased amount of elastic fibers in upper and mid dermis, with increased dermal pigmentation from both melanin and hemosiderin.

Differential Diagnosis

The elastic fibers are intact as opposed to the production of elastotic material seen in solar elastosis.

Hemosiderosis/Hemochromatosis

Clinical

Clinically there is hyperpigmentation of the skin, particularly in exposed areas, that appears brown to bronze in color.

Histology

Histologically, hemosiderin deposits (Perl's stain) appear as fine granules around vascular and adnexae basal membranes, free in the dermis and within macrophages. There is also epidermal melanosis.

Metabolic Pigment Deposits

Ochronosis

Clinical

Endogenous ochronosis is inherited as an autosomal recessive trait, in which homogentisic acid accumulates in cartilage, tendons, sclerae and eventually the skin due to lack of homogentisic acid oxidase. Exogenous ochronosis follows the application of bleaching creams containing hydroquinone.

Histology

Ochronotic pigment (yellow-brown: ochre) accumulates in collagen bundles in upper and mid dermis. The collagen fibers appear swollen, homogeneous and with sharp edges or tapered ends. Fine granules of pigment can be seen throughout the dermis, basal membranes, endothelial cells and secretory cells of sweat glands. The pigment becomes black when stained with cresyl violet or methylene blue.

Differential Diagnosis

The histologic findings are characteristic. Occasionally, melanocytic nevi show upper dermal deposits of brown pigment similar to those seen in ochronosis.

Exogenous Pigment Deposits

Tattoo

Clinical

Intentional tattoos are produced by introducing pigments such as cinnabar, chrome green, and cobalt blue, into the upper dermis. Inflammatory reactions occasionally follow. Accidental tattoos introduce foreign material in the skin, which can occur at the time of an abrasion or a close range fire arm shot.

20

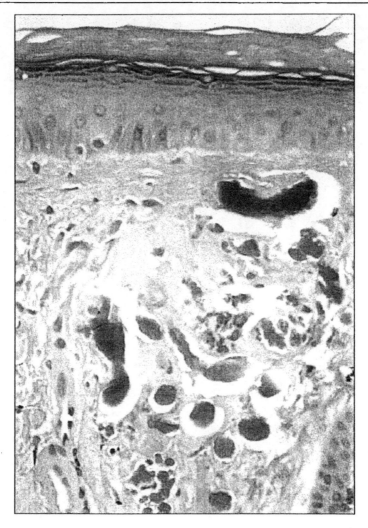

Fig 20.2. **Exogenous ochronosis:** The collagen bundles in the upper and mid dermis are irregularly coated by brownish amorphous material.

Histology

The pigment is seen as irregular granules of dark material free in the dermis and within macrophages. Inflammatory reactions, if present, can be granulomatous.

Differential Diagnosis

Argyria due to prolonged ingestion or topical application of silver salts is characterized by a slate blue discoloration of the skin or mucous membranes. Silver salts

20

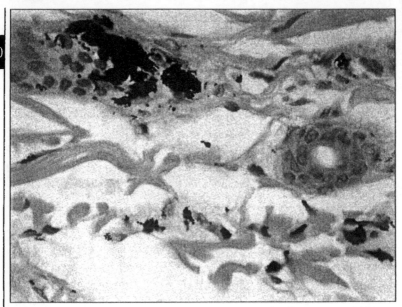

Fig 20.3. **Tattoo**: Granular deposits of black material are seen in the proximity of the superficial capillary plexus. The material is both extracellular and in macrophages.

are occasionally deposited as a result of dental work. Histologically silver deposits, regular in size, are seen in the dermis and the basement membrane of sweat glands. Usually extracellular, they are visualized better with dark field examination. Chrysiasis occurs after prolonged treatment with gold salts, such as for rheumatoid arthritis. The skin shows a bluish discoloration. Gold granules, that are larger than silver granules, are usually found intracellularly, within macrophages and endothelial cells.

Section 56. Melanocytic Tumors

Benign Tumors

Melanocytic nevi are either congenital or acquired. Most acquired nevi develop in the first two decades of life, although crops of new nevi may appear later, and new lesions may be seen throughout life. Nevus cells are derived from melanocytes, but appear rounded instead of dendritic. They are often grouped in nests displaying round to oval nuclei, inconspicuous nucleoli, and ill defined cytoplasm. Mitosis are seen only very rarely in benign nevi, and if present they are always typical. Acquired nevi originate from the basal layer. Lentigo simplex may be a precursor lesion of junctional nevi.

Congenital Nevi

Clinical

They are present at birth or develop within the first year of life. Congenital nevi are usually larger than acquired ones and they can reach very large sizes. Giant congenital nevi (GCN) are at least 18 cm in diameter in the adult patient, smaller in newborns since they grow at the same rate as the body. Hairy nevus refers to the fact that they often have abundant terminal hairs. GCN are often located on the scalp and on that location they often involve meninges. Other common locations are the trunk or the extremities. The chances that a melanoma will develop in a congenital nevus increase with the size of the nevus. Malignant transformation of a GCN may happen early in childhood or during the first decade, therefore surgical treatment of large nevi, if attainable, should be attempted early. Tumor nodules sometimes develop in GCN, which may indicate malignant transformation.

Histology

Congenital nevi often display sheets of nevus cells with small nuclei and indistinct cytoplasm that may resemble lymphocytes (type C nevus cells). They are located throughout the dermis, often seen as deep as the subcutaneous fat. The nevus cells often surround or infiltrate cutaneous adnexae, particularly hair follicles (Fig. 20.4). They may also infiltrate vessel walls or cutaneous nerves. Indian file arrangement is commonly seen. The nodules arising in GCN show cells that are larger, and display irregular nuclei with prominent nucleoli, difficult to distinguish from malignant cells.

Lentigo Simplex/Solaris

Clinical

Lentigo simplex appear as hyperpigmented macules without predilection for sun exposed areas, all ages. Lentigo solaris are hyperpigmented macules on sun exposed areas, middle age to old individuals.

Histology

There is hyperpigmentation along the basal layer with some degree of melanocytic hyperplasia. The latter may be difficult to evaluate if the melanocytes are very pigmented. Small nests of melanocytes may be seen. Lentigo solaris display elongated rete ridges and solar elastosis, features not so prominent in lentigo simplex (Fig. 20.5).

Differential Diagnosis

Nevus spilus, which appears speckled clinically, histologically shows features of both lentigo simplex and junctional nevi, the latter in the darker areas. Multiple lentigines syndromes include the lentiginosis profusa and the multiple lentigines or leopard syndrome.

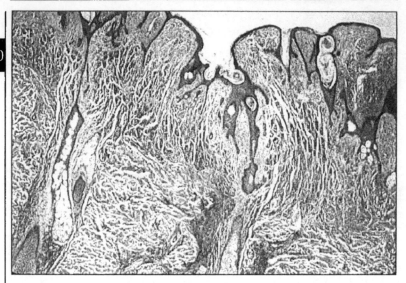

Fig 20.4. **Congenital nevus**: Large compound nevus with nests of small, regular nevus cells infiltrating the reticular dermis and around hair follicles.

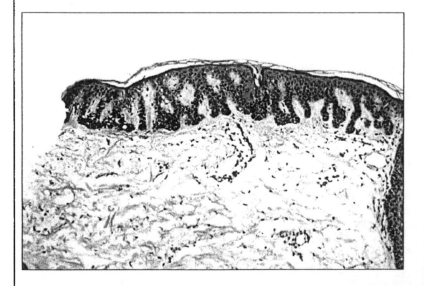

Fig 20.5. **Lentigo solaris**: The epidermis shows elongation of the rete-ridges with hyperpigmentation and hyperplasia of melanocytes. Solar elastosis in the dermis.

Acquired Melanocytic Nevi

Clinical

Junctional, compound, and intradermal nevi are stages in the evolution of a nevus manifested clinically as hyperpigmented macules (junctional), thicker hyperpigmented papules (compound), and larger, dome-shaped papules (intradermal). The latter are of variable pigmentation but they are often nonpigmented. Acquired nevi are usually less than 1cm in diameter, and stop growing after they become intradermal (resting phase). Nevi may appear anywhere in the body, although they have a predilection for the trunk, head and neck. Acquired melanocytic nevi appear sometimes as fibroepithelial polyps.

Histology

Junctional nevi show nests of melanocytes usually located on the tips of the rete ridges. They are often hyperpigmented, which often results in pigment incontinence with melanophages seen in the papillary dermis. The nevi cells at the dermoepidermal junction appear larger, with abundant cytoplasm (type A nevi cells). Compound nevi show nests of nevus cells both at the dermo-epidermal junction as well as in the dermis (Fig. 20.6). The dermal component may be minimal or prominent. Intradermal nevi show most of the nevus cells in the dermis. They are usually arranged in nests with ill defined cytoplasmic boundaries, which make them difficult to distinguish from lymphocytes. The later is particularly true in the lower part of the nevi, since there is a maturation phenomena, which explains the fact that cells are smaller with less cytoplasm in the deeper part of the nevus. Intradermal nevi expand the papillary dermis, with little if any involvement of the reticular dermis, in contrast to congenital nevi. Neurotized nevus shows nests of nevi cells that resemble nerve corpuscles.

Recurrent Nevus

Clinical

Compound or intradermal nevi that have been incompletely removed by shave biopsy may recur. Clinically a hyperpigmented macule appears on the site of a previously biopsied nevus.

Histology

Histologically recurrent nevi show junctional activity characterized by nests of melanocytes that may appear atypical and are often quite pigmented. The upper dermis shows a scar, under which a residual intradermal nevus is present. Beware, recurrent nevi may resemble melanoma.

Halo Nevus

Clinical

Clinically appear as nevi surrounded by a hypopigmented halo. Occasionally seen in patients with previous melanoma, but is usually a nonsignificant finding, often seen in children.

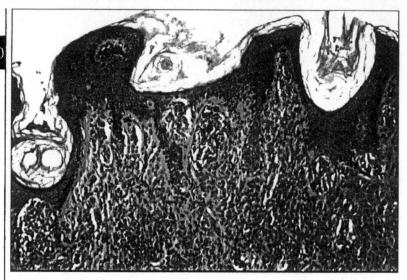

Fig 20.6. **Compound nevus**: Nevus cells are seen as nests at the dermo-epidermal junction, and as sheaths in the dermis.

Histology

Histologically halo nevi are junctional or compound nevi in which there is a marked lymphocytic response. The infiltrate is lichenoid, reaching and obscuring the basal membrane zone. The nevus cells are often difficult to see among the lymphocytes and eventually end up disappearing. Some inflamed nevi do not show a clinical halo. The differential diagnosis includes benign lichenoid keratosis, and fixed drug eruption.

Balloon Cell Nevus

Histology

Occasionally intradermal or compound nevi histologically show balloon cell changes, characterized by large, round and clear cytoplasm, with small centrally located nucleus. The balloon cells are two or three times the diameter of regular nevi cells. Special stains for glycogen or mucosubstances in those cells are negative. Although small foci of balloon cell degeneration is relatively common, when the proportion of balloon cells exceeds 50% of nevi cells it is proper to designate it as a balloon cell nevus.

Differential Diagnosis

The main differential diagnosis is with balloon cell melanoma, which in addition to the balloon changes also show atypical features.

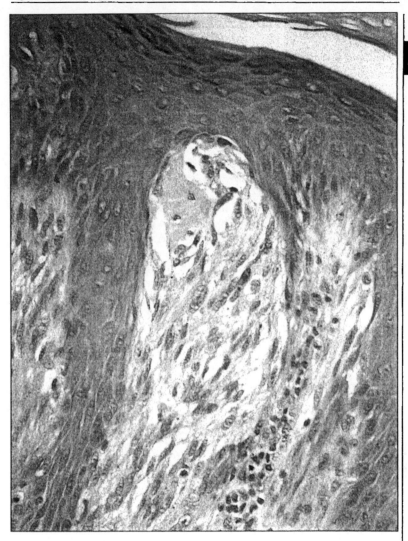

Fig 20.7. **Spitz nevus**: The dermal papillae are filled with large spindle nevus cells. An eosinophilic, homogeneous Kamino body is seen at the dermo-epidermal junction.

Spitz Nevus: Spindle and Epithelioid Cell Nevus

Clinical

They appear as rapidly growing red papules in children and young adults. Be very cautious about diagnosing Spitz nevi in patients over forty years of age.

Histology

Spitz nevi are usually compound nevi (junctional and intradermal Spitz nevi may also occur). The architecture is symmetrical and characterized by an inverted triangle with the apex in the deeper part of the lesion. Spitz nevi SHOULD NOT HAVE lateral junctional extension (shoulder) or intraepidermal invasion (Pagetoid spread). At the dermoepidermal junction there are large nests that exhibit focal separation from the epidermis (capping). At the epidermal/melanocyte interface necrotic, eosinophilic cells (apoptotic?) may be seen, i.e., Kamino bodies. The nevus cells often show large, round to oval cytoplasm (epithelioid cells), or else they are spindle. The nuclei are enlarged and they may be mild to severely atypical (Fig. 20.7). They usually have a prominent nucleolus that is eosinophilic, in contrast to the purplish nucleolus of melanoma cells. Mitosis may be present, particularly in the most superficial areas, but atypical mitoses should be absent. Spitz nevi often exhibit some degree of maturation, with the cells in the lower part of the lesion being smaller and less cohesive (rain drops). They are often amelanotic, but presence of melanin would not exclude the diagnosis of Spitz. A lymphocytic infiltrate is not characteristic of these lesions, particularly if it penetrates among the nevi cells. Necrosis would suggest melanoma.

Pigmented Spindle Cell Nevus of Reed

Clinical

Clinically they appear as flat, very pigmented nevi, most common on the lower extremities of young adults.

Histology

Histologically they are junctional nevi in which the melanocytes are spindle and display a plate-like growth, parallel to the surface, with bridging of the rete ridges. There is abundant melanin both intracellular and as pigment incontinence. These lesions are considered by many as junctional variants of Spitz nevus.

Blue Nevus, Common

Clinical

Small, usually less than 1cm., round macules with a dark blue color, often on extremities. They may be congenital or acquired.

Histology

Elongated, dendritic melanocytes are seen in the dermis. They can be located at all levels, sometimes very close to the epidermis, other times in the deep dermis. They are usually pigmented, and the melanin granules in the cytoplasm of the cells allow visual-

20

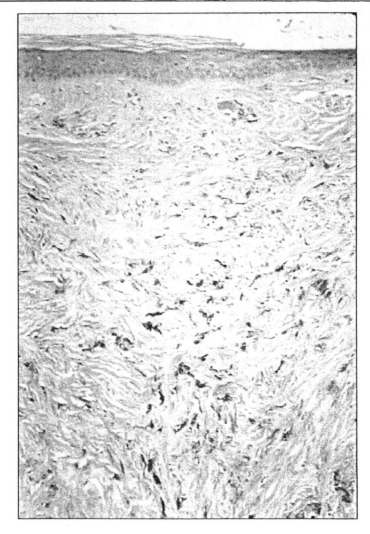

Fig 20.8. **Common blue nevus**: In the dermis there is an infiltrate of spindle, dendritic melanocytes, many of which are pigmented.

ization of the dendritic processes. Melanophages are also seen around the tumor cells. The tumor cells are often seen around hair follicles (periadnexal dermis) (Fig. 20.8).

Differential Diagnosis

Nevus of Ota and Ito exhibit a brown mottled discoloration and are present at birth, or appear in the first two decades of life. The nevus of Ota is located on the

20

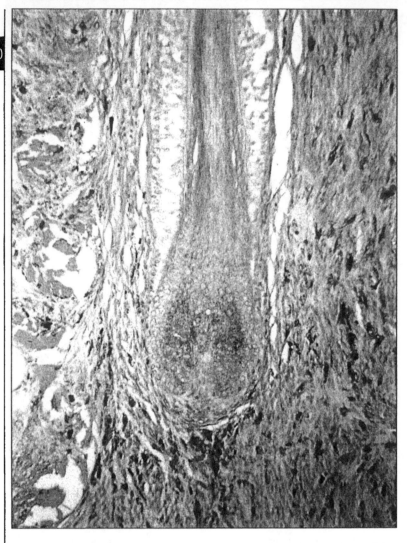

Fig 20.9. **Cellular blue nevus**: Surrounding a hair follicle there is a highly cellular infiltrate of spindle melanocytes, some of which are pigmented.

face and forehead, unilateral or bilateral. The nevus of Ito is located on the shoulder, and the clavicular or scapular regions. Histologically they show scattered pigmented dendritic melanocytes usually seen in the superficial half of the dermis. Mongolian Spot is present at birth, more common in oriental and black children, on the sacrococcygeal region. It appears as an area of bluish discoloration. Histologically, dendritic melanocytes containing melanin granules are seen in the lower part of the dermis. The number of melanocytes in blue nevi is usually considerably larger than in these three entities.

Blue Nevus, Cellular   20

Clinical
Dark, blue papules, usually 1 to 3 cm. in diameter, often located on the buttocks or sacral region.

Histology
Cellular aggregates of large spindle cells are seen in the dermis, often with a wedge configuration that suggests a perifollicular arrangement. The cells have oval nuclei and abundant cytoplasm and are closely placed, with little intervening stroma in between. They are often nonpigmented and form bundles of cells. Cellular atypia is sometimes seen. Pigmented dendritic cells of common blue nevus can be seen also in other areas of the tumor (Fig. 20.9).

Differential Diagnosis
Atypical cellular blue nevi are distinguished from melanoma by the absence of necrosis and little lymphocytic infiltrates. Atypical mitoses are not seen in cellular blue nevus. Deep penetrating nevi are seen in the first four decades, usually less than 1 cm. in diameter and clinically often show variegation in color. Histologically they have a wedge shape, and involve the reticular dermis and occasionally the subcutis. Spread of the lesion along neurovascular tracks often gives a dispersed appearance. The melanocytes are fusiform or epithelioid, display mild to severe atypia, and mitosis may be present.

Premalignant And Malignant Tumors

Dysplastic Nevus (Nevus with Architectural Disorder and Melanocytic Atypia)

Clinical
Junctional or compound nevi with irregular borders, evidence of lateral extension and different shades of color are suspected of being dysplastic. They may have a dot-like area of hyperpigmentation. Dysplastic nevus syndrome refers to patients with hundreds of nevi, many of which are irregular i.e., dysplastic, and with family history of melanoma. Those patients are at high risk for developing melanoma, and should be followed closely. It is not unusual to see remnants of dysplastic nevus in superficial spreading melanomas, even in patients without the dysplastic nevus syndrome, pointing out the relationship between the lesions. However, the presence of one or few dysplastic nevi in a patient without the syndrome should not be construed as increased risk to develop melanoma, as most patients have some nevi with a degree of dysplasia.

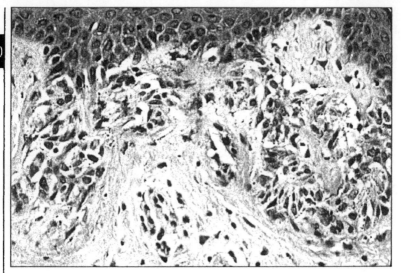

Fig 20.10. **Dysplastic nevus**: There is a proliferation of nevus cells at the tips of the rete, with bridging, cellular atypia, dermal fibrosis, inflammation and pigment incontinence.

Histology

There are degrees of dysplasia in different nevi, with some being mild and focal, and others being severe and approaching melanoma in situ. There are two components to a dysplastic nevus, architecture and cytology. The irregular architecture is the result of junctional melanocytic activity, melanocytic hyperplasia, in a pre-established nevus. The melanocytic hyperplasia may be focal and usually extends laterally beyond the confines of the previous nevus producing a shoulder. The newly formed melanocytes are pigmented and are seen either in nests near the tips of the rete ridges, or as individual melanocytes proliferating along the basal layer, lentiginous hyperplasia. There is usually bridging of the rete ridges by the proliferating melanocytes. The nevus cells show cytologic atypia, characterized by large nuclei, presence of nucleoli and abundant smoky cytoplasm. An attempt should be made to grade the cellular atypia. Contributing to the architectural disarray is the host response, characterized by a lymphocytic infiltrate which results in pigment incontinence and fibrosis of the dermis. The later is one of the most significant findings in a dysplastic nevus and it is characterized by layered, newly formed fibrous tissue around the rete ridges (Fig. 20.10). It should be distinguished from the hyalin fibroplasia surrounding the rete ridges which is seen not only in dysplastic nevus, but to an extent in lentigo simplex/solaris and in common junctional nevi. The fibrosis of the dermis may play a role in angulating the nests of melanocytes and orienting them parallel to the surface. In summary there is a host response in dysplastic nevi which is responsible for many of the changes found and that can be compared to the phenomenon of regression in melanoma. We diagnose dysplastic nevus as "junctional (or compound) nevus, dysplastic pattern, minimal (or moderate, severe) atypia". In the case of severely dysplastic nevi we recommend limited re-excision.

20

Melanoma

Melanoma denotes a malignant tumor of melanocytes with the capability to invade the surrounding tissues and produce regional and distal metastases. As with tumors in other areas, we recognize a stage that we call in-situ and an invasive stage. Since we base our diagnosis on morphologic grounds, it is not very clear whether the cells of melanoma in-situ have the capability of independent growth as defined above, or when and why they become malignant, i.e., invasive. The in-situ melanomas are said to be in radial growth phase, while the invasive melanomas are in vertical growth phase. The radial growth phase is lateral superficial growth of melanoma along the epidermis, and perhaps the papillary dermis, without invading the reticular dermis. Lentigo maligna and superficial spreading melanoma may remain in the radial growth phase for a long time without deep invasion. Vertical growth phase denotes invasion. Nodular melanoma usually expresses this type of growth from the beginning, without lateral, superficial spread. Acral lentiginous melanoma usually displays both types of growth early on. While the above four types of melanoma differ in how rapidly they become invasive, once a melanoma starts to invade, the prediction of the biologic behavior of that tumor is based in other parameters unrelated to the type of melanoma from which it originated.

Lentigo Maligna

Clinical

Usually seen on the face of older patients, it can be seen also on the arms and upper trunk. It presents as an irregularly shaped, highly pigmented macule that enlarges gradually to reach a diameter often in excess of 3 or 4 cms. If regression has taken place, the appearance is mottled, with different shades of brown, red, black and even depigmentation.

Histology

There is atrophy of the epidermis, which appears flat with disappearance of the rete ridges. Along the basal layer is a lentiginous (individual cell) proliferation of spindle melanocytes with vacuolated cytoplasm, usually heavily melanized. The atypical melanocytes involve also the basal layer of the hair follicles, a finding helpful in the diagnosis of lentigo maligna. In long standing lesions nests of melanocytes may be seen along the basal layer, although pagetoid invasion of the epidermis is not present (Fig. 20.11). The upper dermis shows solar elastosis as well as pigment incontinence. Long standing lesions eventually will develop invasion of the dermis at one or more foci. When the lesion becomes invasive (vertical growth phase) it is designated lentigo maligna melanoma, and then behaves as other invasive melanomas (see below).

Differential Diagnosis

The histologic findings in peripheral areas of a lentigo maligna may not show atypical melanocytes and therefore cannot be differentiated from lentigo solaris with

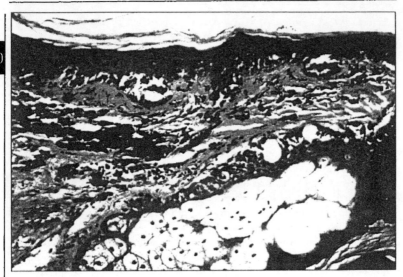

Fig 20.11. **Lentigo maligna**: A lentiginous proliferation of melanocytes at the basal layer, and along the hair follicle. The epidermis is atrophic and there is pigment incontinence.

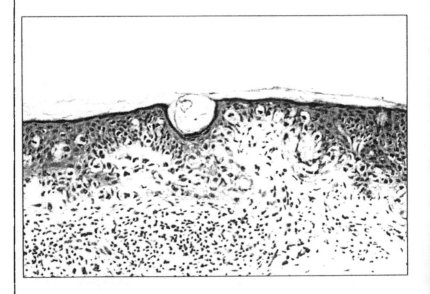

Fig 20-12. **Superficial spreading melanoma**: Clear, atypical melanocytes are growing in nests at the D-E junction and infiltrating the epidermis with a pagetoid spread.

a flat epidermis. If there is clinical suspicion of melanoma, several samples may have to be taken.

Superficial Spreading Melanoma

Clinical

It is the most common type of melanoma, perhaps as many as 70% of all melanomas fall in this category. It is seen more often on the upper back and legs of young to middle age men and women. The lesions are usually over 1 cm. in diameter and present as an irregular macule with different shades of color, ranging from black to brown, red, pink and white. The variegation indicates regression. If a nodule can be palpated invasion may be suspected. This tumor may grow in the area of a previously existing nevus.

Histology

There is hyperplasia of atypical melanocytes that are forming nests along the dermo-epidermal junction, usually near the tips of the rete ridges. Skip areas without tumor cells are frequently seen, with other tumor nests growing two or three rete ridges apart from the main lesion. The atypical melanocytes exhibit large round to oval vacuolated nucleus with prominent eosinophilic nucleolus. Intranuclear cytoplasmic vacuoles are often seen. Mitoses are generally present, and they are sometimes atypical (polyploid). The cytoplasm of the tumor cells is usually clear with a variable amount of melanin deposition. A pathognomonic finding is the invasion of the epidermis by individual or small nests of tumor cells. The pagetoid spread of the epidermis is diagnostic for this type of melanoma (Fig. 20.12). The upper dermis shows a lymphocytic infiltrate and pigment incontinence. Regression implies a host response and is manifested histologically by fibrosis of the dermis, dilated capillaries and melanophages. Regression should be reported even in the absence of dermal invasion, since it may indicate a previously invasive tumor and it worsens the prognosis. If there is invasion of the dermis the tumor should be considered invasive regardless of its origin in a superficial spreading melanoma.

Nodular Melanoma

Clinical

They are sharply delimited nodules with a variable degree of pigmentation ranging from black to light brown or totally amelanotic. In the last instance they may be red and mistaken for hemangiomas. They start as papules that grow rapidly and ulcerate. Occasionally they may be pedunculated. Only about 10% of all melanomas are nodular but they have the highest mortality rates.

Histology

There is a proliferation of atypical melanocytes that are invasive from the inception (vertical growth phase) with sharply delimited lateral borders. The tumor cells may infiltrate and ulcerate the overlying epidermis, and they usually invade the reticular dermis by the time the tumors are biopsied. The tumor cells are of large size, although they may be of deceptively regular size and shapes (Fig. 20.13). Other

20

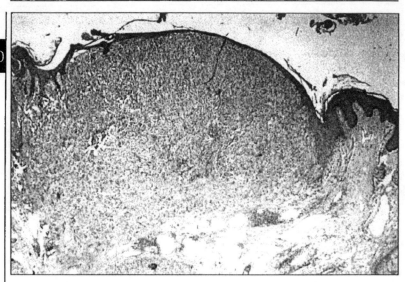

Fig 20.13 . **Nodular melanoma**: Nodule of atypical melanocytes filling up most of the dermis without radial, superficial extension. Abundant mitosis, and dysplastic features.

times they may show marked anisocytosis and atypia. There is usually a large, eosinophilic nucleolus, and intranuclear cytoplasmic inclusions are often seen. The cytoplasm is abundant and it may be totally depigmented even after Fontana stain. Multinucleated and/or giant cells may be seen. Mitoses are frequent and it is not uncommon to see atypical mitoses. There is usually a lymphocytic infiltrate surrounding the tumor and penetrating among the tumor cells. Areas of necrosis may be seen. Tumor cells often involve cutaneous adnexae or blood vessels.

Acral Lentiginous Melanoma

Clinical
This type of melanoma is found on the palms, soles, nail matrix and mucous membranes. It is the most common type of melanoma in orientals, blacks, and other individuals with darker skin. It usually has a poor prognosis.

Histology
There are large, dendritic melanocytes along the basal membrane, seen as individual cells or forming small nests. The epidermis is hyperplastic and shows elongation of the rete ridges. These tumors exhibit both radial growth phase and vertical growth phase from the beginning.

Invasive Melanoma
Nodular melanoma is invasive from inception. Acral lentiginous melanoma is often invasive by the time it is diagnosed. Lentigo maligna and superficial spreading

melanoma, on the other hand, are often found in the radial growth phase. Early invasion in those two tumors is manifested by nests of cells, similar to those found in the epidermis, invading the papillary dermis (unconnected to the epidermis). More advanced invasion shows tumor nests reaching the superficial capillary plexus, filling up and expanding the papillary dermis or penetrating the reticular dermis or beyond. In those cases we report our findings as: "Invasive melanoma, arising in a superficial spreading melanoma (or Lentigo maligna melanoma), Clark's level __, __ mm. in maximum thickness, margins free (or involved)". In a note in the report, other prognostic factors such as ulceration, lymphocytic infiltrate, mitoses, presence or absence of regression etc., can be mentioned. We usually recommend re-excision. A word of caution: a preexisting intradermal nevus is sometimes present in the area of melanoma. Be careful when measuring thickness of invasion not to include a benign nevus as part of the tumor.

Prognostic Factors in Melanoma

Levels of Invasion and Tumor Thickness
The prognosis of a melanoma worsens as the depth of invasion increases. To measure invasion the two most commonly used parameters are Clark's levels of invasion and Breslow's tumor thickness. The levels of dermal invasion according to Clark are:

> Level 1—intraepidermal;
> Level 2—invasion of dermal papillae;
> Level 3—involvement of superficial vascular plexus (filling up the papillary dermis);
> Level 4—invasion of reticular dermis;
> Level 5—invasion of subcutaneous tissue.

The maximum thickness of the tumor is measured from the granular layer (or equivalent in the case of ulceration) to the deepest part of the tumor. Traditionally very thin melanomas are those that measure less than 0.76 mm, thin melanomas measure 0.76-1.5 mm, intermediate melanomas measure 1.5-3.0 mm, and thick melanomas measure over 3.0 mm. The prognosis for very thin and thin uncomplicated melanomas is over 90% five year survival. However, other prognostic factors can modify the outcome, including ulceration of the epidermis, presence of regression, number of mitoses, lymphocytic infiltrate, invasion of vessels, necrosis, and lymphatic metastases.

Staging of Melanoma
Staging follows the TNMM classification. Stages I and II represent local disease. Stage III regional lymph node and/or in-transit disease. Stage IV distant disease.

Treatment of Melanoma
The main form of treatment is still surgical. A limited re-excision of about 1 cm, is recommended for thin melanomas, while thick melanomas are re-excised with 2-3 cm. margins. Prophylactic regional lymph node dissection often is done in melanomas of intermediate thickness. An alternative is the sentinel node biopsy. Chemotherapy and immunotherapy have been used in the past without great success.

Presently interferon alpha is used as an adjuvant to the surgical treatment in thick melanomas or metastatic melanoma.

Other Types of Melanoma

Desmoplastic and Neurotropic Melanoma

Seen in older patients on the face or upper extremities. Histologically is characterized by spindle cells without marked atypia, and fibrosis of the dermis. A junctional component may be absent. It may be confused with dermatofibroma. Sometimes these tumors infiltrate and spread through nerves, i.e., neurotropic melanoma.

Balloon Cell Melanoma

It is the malignant counterpart to balloon cell nevus. Usually in addition to balloon cells other epithelioid melanoma cells are also present.

CHAPTER 21

Vascular and Neural Tumors

Ramón L. Sánchez and Sharon S. Raimer

Section 57. Vascular Tumors and Malformations

There are multiple vascular proliferations involving the skin, most of them benign. Since their morphologies are variable, the classifications are not totally satisfactory. It is helpful to divide the vascular proliferations into tumors and malformations. Both of them can be congenital or acquired. Furthermore, they should be classified as superficial and deep, the latter involving full dermis and at least some of the subcutaneous tissue. Most vascular malformations will have a deep component in addition to the superficial one. Vascular tumors, on the other hand, are often superficial, particularly in the case of most acquired capillary hemangiomas.

Portwine Stains/Nevus Flameus

Clinical

Vascular malformations are usually present at birth, or shortly thereafter. They are characterized by large macular purplish areas (portwine), usually located on the face or elsewhere. With time they may become thicker vascular plaques. On the face they may be limited to the cheeks or involve also the ocular and frontal areas. The latter location is frequently associated with the Sturge-Weber syndrome. In Sturge-Weber syndrome there is a facial port-wine stain with leptomeningeal involvement underlying cerebral atrophy and calcification with epilepsy. Extensive port-wine stains on the extremities may be a sign of Klippel-Trenaunay-Weber syndrome. In Klippel-Trenaunay-Weber syndrome arteriovenous malformations resulting in hypertrophy of soft tissue and bone of an extremity are associated with an extensive nevus flammeus.

Histology

Histologic features are deceptive and characterized exclusively by a slightly increased number of blood vessels in the upper and papillary dermis, most of them being dilated telangiectasias.

Capillary Hemangiomas, Congenital

Clinical

Usually apparent at birth or shortly thereafter, they often involve the head, although they can also be seen on extremities. They are clinically characterized by papules, nodules, tumors with a red (strawberry) appearance. The size varies; some

---

Dermatopathology, edited by Ramón L. Sánchez and Sharon S. Raimer. ©2001 Landes Bioscience.

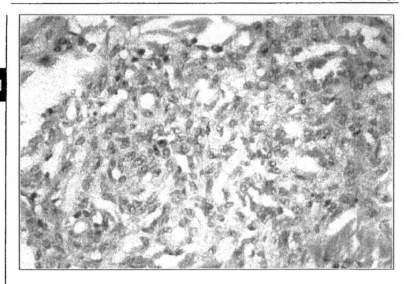

Fig 21.1. **Capillary hemangioma**: Proliferation of capillaries and endothelial cells forming solid tumor lobules.

are quite large. They may ulcerate superficially. They are often superficial, although some capillary hemangiomas can be quite deep, involving subcutaneous tissue, as well as underlying muscle. A majority of superficial capillary hemangiomas resolve spontaneously within the first decade of life, but some hemangiomas, particularly the deeper ones, may persist throughout life.

Histology
There is a proliferation of capillaries, as well as endothelial cells, which usually present a lobular arrangement. The proliferation of endothelial cells can be so extensive as to obscure the presence of small lumina (Fig. 21.1). Some of the endothelial cells may show typical mitosis. When there is superficial ulceration, an acute infiltrate, particularly of PMNs, is seen.

Capillary Hemangiomas, Acquired

Pyogenic Granuloma (Lobular Capillary Hemangioma)

Clinical
They present as red often pedunculated, papules, usually less than one centimeter in diameter, often in areas of previous trauma. A small collarette may surround the base of the lesion. A characteristic location is on the oral mucosa during pregnancy. They are often superficially ulcerated, and occasionally they may develop satellite lesions.

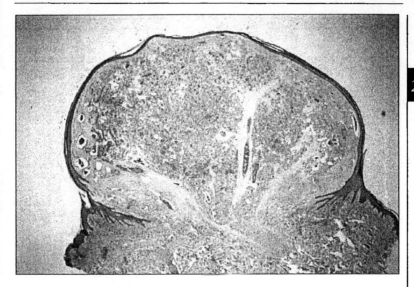

Fig 21.2. **Pyogenic granuloma**: Exophytic lobular proliferation of capillaries and endothelial cells. Notice the collarette at the base of the lesion.

Histology

Well-delineated papule covered by epidermis, with a collarette on the area of attachment to the adjacent uninvolved skin. Inside the papule there is a capillary proliferation with small lumina and a proliferation of endothelial cells with a lobular arrangement (Fig. 21.2). Since superficial ulceration is the rule, an acute inflammatory infiltrate is usually present. The stroma is usually edematous. The tumor often involves adjacent reticular dermis.

Other Capillary Hemangiomas

A number of capillary hemangiomas that can be characterized by their clinical and/or histologic appearance have been identified.

Histologically, two main subtypes can be differentiated. Some of them are paucicellular, characterized by dilated capillaries with few endothelial cells, while others are multicellular and characterized by a proliferation of endothelial cells, with only small lumen. A combination of both can sometimes be seen. Among the first type are **cherry angiomas**, very common tumors on middle-aged and older patients, clinically seen as one or multiple small bright red to purple papules, on the trunk and extremities. Histologically, a proliferation of dilated capillaries is seen. **Tufted angiomas**, on the other hand, are acquired vascular tumors of infancy and childhood on the neck and trunk or extremities that can be quite large. Histologically there is a proliferation of endothelial cells with small lumen, which are arranged in well-defined round cellular aggregates scattered throughout the superficial and reticular dermis, that have been compared to cannonballs.

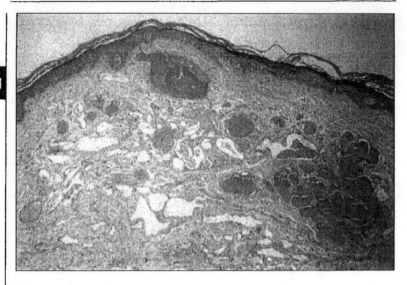

Fig 21.3. **Cherry angioma**: Dilated capillaries in the papillary dermis without endothelial cell proliferation.

Targetoid Hemosiderotic Hemangioma

Clinical
Shows an annular, target-like clinical appearance.

Histology
There are two different components. A superficial component near the papillary dermis is made of dilated capillaries. In contrast the mid and deep dermis show a proliferation of endothelial cells and small lumen that appear to be infiltrating and dissecting between the collagen bands suggesting Karposi's sarcoma. However, the typical clinical appearance seen in the setting of a nonimmunocompromised young patient, and histologically the absence of atypia, and of true invasion, allow the recognition of this entity.

Cavernous Hemangiomas
Traditionally, this term refers to vascular tumors characterized by large dilated vascular spaces which are often seen not only in the superficial dermis, but also in the deep dermis and the subcutaneous tissue (Fig. 21.4). Some of these tumors can be very deep, infiltrating muscle or fascia. By extension, the term cavernous is applied to large dilated vascular spaces. For instance, a vascular tumor that is mostly cellular with focal dilated vascular spaces is said to have a "cavernous component." Possibly cavernous hemangiomas are mainly formed by veins, and some of them may be vascular malformations, rather than neoplasms. Likewise some tumors diagnosed as cavernous hemangiomas can be arteriovenous malformations.

21

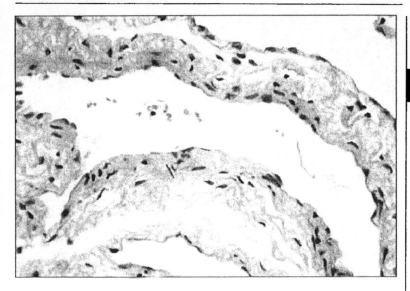

Fig 21.4. **Cavernous hemangioma**: Large vascular spaces, often involving deep dermis and even subcutaneous tissue.

Arteriovenous (AV) Malformations and Hemangiomas

Clinical

The typical A-V malformation is either congenital, or acquired, growing spontaneously or secondarily to trauma. An A-V malformation usually implies defective and abnormal anastomosis between an artery and a vein. Sometimes a bruit can be heard on ausculation.

Histology

The dermis shows a vascular proliferation characterized by vessels with different wall thicknesses, some of them with all of the layers seen in the wall of arteries, while others are typical veins. Some other vessels show a combination of venous and arterial histologic features. The actual arteriovenous anastomosis is often deep, and cannot be identified. The proliferation of vessels seen can be quite pronounced, and sometimes it can suggest the diagnosis of a malignant vascular tumor, particularly Kaposi's sarcoma. This is seen with some frequency on the ankles or feet of some patients in an entity that is designated **acroangiodermatitis, or pseudo-Kaposi's Sarcoma**. The key to the diagnosis of this entity is the lobular arrangement of the capillary and blood vessel proliferation, as opposed to the invasive, dissecting nature of true Kaposi's as well as the lack of atypical features seen in acroangiodermatitis (Fig. 21.5). In some patients with vascular tumors in which the histology shows a mixture of mid-sized veins and arteries, and in which no arteriovenous malformation has been identified, the diagnosis of **arteriovenous hemangioma** has been used. Another entity

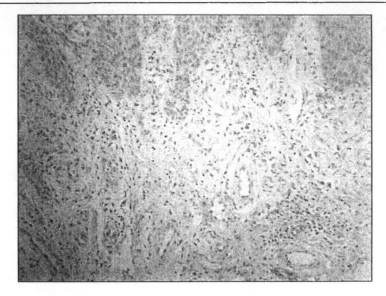

Fig 21.5. **Acroangiodermatitis**: Vascular proliferation throughout the dermis with a lobular arrangement. Differential diagnosis with Kaposi's sarcoma should be established.

that presents with an arterio-venous mixture of blood vessels is the **blue rubber bleb syndrome**, which usually presents at birth, although it becomes more evident in the second or third decades, characterized by large protruding vascular malformations throughout the body and internal organs such as the gastrointestinal tract. Histologically blue rubber bleb syndrome shows large dilated vascular spaces with different wall thicknesses.

Telangiectasias

Clinical

Telangiectasias are vascular disorders that are more noticeable clinically than histologically. They are often seen on the face or on the extremities and sun exposed areas in middle-aged and older patients. **Progressive essential telangiectasia** is a congenital condition characterized by increased involvement of large areas of the body by telangiectasia. **Essential familial telangiectasia** (Rendu-Weber-Osler) is clinically characterized by multiple matt-like telangiectasias on the lips, oral mucosa and fingertips. These patients relate a history of frequent nasal bleeding. **Vascular spiders**, or nevus aracneous, are mainly seen on the face or trunk, and are characterized by a central feeding blood vessel from which telangiectatic blood vessels arise in a whorled spider configuration.

Histology

Telangiectasias are seen histologically as dilated blood vessels or capillaries in the upper dermis. In nevus aracneous, if the biopsy includes the feeding vessel, a small artery is seen, usually with an irregular course, surrounded by dilated capillaries.

Venous Lakes

Clinical

They are seen mainly in elderly people on the lips or other sun exposed areas. They are characterized by a single bluish papule that can be compressed, and that should be differentiated clinically from a blue nevus, or other melanocytic lesions.

Histology

A large vascular dilated space is seen.

Glomus Tumor

Clinical

The typical Glomus tumor appears in the nail bed, near the proximal fold of the nail, and is painful. Solitary Glomus tumors, however, can be seen anywhere. Multiple Glomus tumors have no special predilection for a single area.

Histology

Solitary Glomus tumors may be encapsulated and are characterized by blood vessels surrounded by several layers of glomus cells. The tumor cells are arranged concentrically around the lumen, immediately outside the vessel walls (Fig. 21.6). They are small homogeneous cells with scanty cytoplasm. In contrast, multiple Glomus tumors, or glomangiomas, exhibit dilated vascular spaces surrounded by only two or three layers of glomus cells, which are arranged in an orderly fashion (Fig. 21.7). Glomangiomas are not encapsulated tumors.

Stains

Glomus cells are contractile cells in charge of closing or opening arteriovenous anastomoses in the canals of Suquet-Hoyer. As such, they have cytoplasmic filaments that stain positive for smooth muscle actin. Glomus cells are also positive for vimentin.

Angiokeratomas

Clinical

This group of vascular tumors can be congenital or acquired. Depending on the presentation, they have been given different eponyms, such as angiokeratoma of Fordyce when located on the scrotum. Similar tumors can be seen also on the vulva. Angiokeratoma of Mibelli are small papules on the dorsa of fingers or toes seen in childhood or adolescence. Other types are angiokeratoma circumscriptum, and multiple angiokeratomas, which are often seen on extremities. Angiokeratoma corporis diffusum refers to the multiple small vascular tumors seen in association with

21

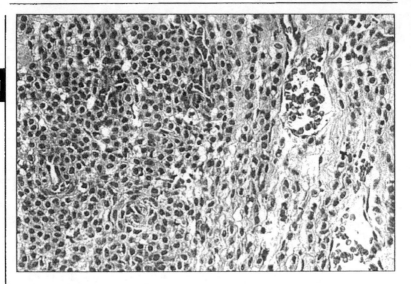

Fig 21.6. **Glomus tumor, solitary**: Vascular lumen surrounded by sheets of rounded glomus cells vaguely arranged in rows. Tumor cells are smooth muscle actin positive.

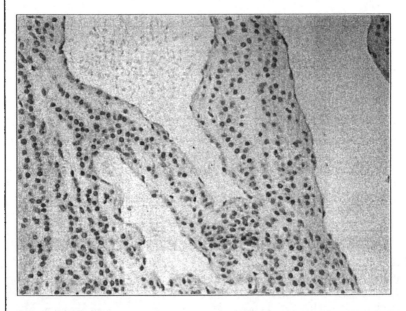

Fig 21.7. **Glomangioma**: Large vascular spaces lined by endothelial cells and surrounded by few layers of glomus cells arranged in layers.

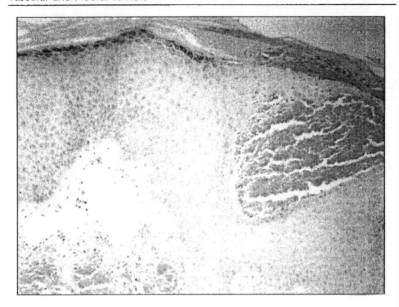

Fig 21.8. **Angiokeratoma**: Vascular spaces in the papillary dermis in close proximity to the epidermis. Some of the spaces appear to be intraepidermal.

Fabry's disease, possibly the result of deposits of lipids in endothelial cells seen in this X-linked genetic condition, in which there is a deficiency of α-galactosidase.

Histology

Angiokeratomas all have in common a proliferation of dilated capillaries in the papillary dermis with some degree of hyperplasia of the epidermis, as well as hyperkeratosis. Because of the proximity of the capillaries to the epidermis and the cut incidence of the biopsy, some of the capillaries appear to be in the epidermis, and this is a diagnostic feature for angiokeratoma (Fig. 21.8). In the case of Fabry's disease, the vascular proliferation is rather small, showing also dilated capillaries in the papillary dermis. Because of the lipid deposits in the cytoplasm of the endothelial cells of the involved capillaries, PAS stain will show positive granules. Lipid stains will also be positive on fresh tissue. Electron microscopy of the affected capillaries shows diagnostic cytoplasmic lysosomal inclusions.

Lymphangiomas

Lymphangiomas can be circumscribed and superficial or can be rather large and ill-defined with superficial and deep components.

Clinical

Superficial well-circumscribed lymphangiomas (lymphangioma circumscriptum) appear as groups of translucent vesicles protruding from the epidermis.

Lymphangiomas frequently have deeper components and may produce enlargement of the organs where they are located, such as the tongue or the neck tissue. Lesions with larger channels have been referred to as cavernous lymphangiomas and very large lesions in which transluminence often can be demonstrated. Translucent vesicles are frequently seen on the surface of deeper lesions.

Histology

Dilated vascular spaces that contain a homogenous or granular pink material (lymph) are diagnostic. Lymphatic valves can be seen however, the lymphatic spaces sometimes are filled with blood, making the differentiation of lymphangioma versus hemangioma virtually impossible on an H&E stain (See note). In the case of diffuse lymphangiomas a recognition of lymphatic spaces in soft tissue or dermis can be difficult, and it often requires a high degree of suspicion.

Note: Immunoperoxidase stain for endothelial cells is usually possible with factor VIII-RA, as well as with CD-34 and CD-31. Vascular endothelial cells stain positive with factor VIII-RA and CD-34, while lymphatic endothelial cells stain positive with CD-34, but not with factor VIII-RA. Theoretically, therefore, it is possible to differentiate lymphangiomas from hemangiomas by means of immunoperoxidase stains.

Angiolymphoid Hyperplasia with Eosinophilia (ALHE)

Clinical

These are vascular tumors that present as nodules that usually are less than one centimeter in diameter, although they may reach several centimeters. A favorite location is on the occipital scalp behind the ears. However, they have been described elsewhere. Often solitary tumors, multiple ALHE tumors have been described on the hands. On this location intravascular involvement of both veins and arteries have been documented. While earlier these tumors were considered to be synonymous with Kimura's disease, they are presently considered separate entities.

While ALHE is a neoplasia, Kimura's disease is considered to be a hypersensitivity reaction.

Histology

The histologic diagnostic feature in ALHE are plump, large endothelial cells that line dilated capillaries. These endothelial cells are considerably larger than normal endothelium. They have regular nuclei and clear cytoplasm, which may sometimes contain phagocytized hemosiderin granules. (Hence the original inclusion of these tumors in the so-called histiocytoid hemangiomas). The capillary proliferation with prominent endothelial cells has a lobular architectural arrangement. Between the lobules there is a prominent infiltrate characterized by lymphocytes, some plasma cells, and a relatively large number of eosinophils (Fig. 21.9).

Note: ALHE is included among the epithelioid hemangiomas, vascular tumors with large, plump endothelial cells, with manifestation not only in the skin, but also

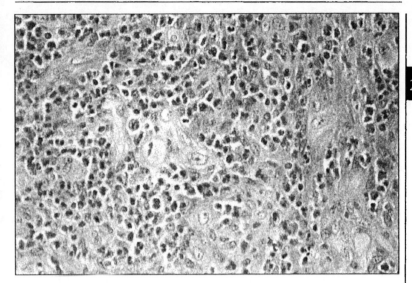

21

Fig 21.9. **Angiolymphoid hyperplasia with eosinophilia:** Vascular proliferation with large histiocytoid endothelial cells and a prominent mixed infiltrate with eosinophils.

in multiple organs. A more aggressive or premalignant form of epithelioid hemangiomas has been termed epithelioid hemangioendothelioma.

There are some vascular tumors that, although benign, resemble malignant vascular formations. Among them is the **spindle cell hemangioendothelioma**, a tumor composed of spindle cells, for the most part, with a cavernous component in the superficial area of the tumor. This tumor can resemble Kaposi's sarcoma, but the spindle cells are more homogeneous and they lack atypical features. Another entity which is composed of atypical vascular formations is **papular angioplasia**, in which there is a proliferation of endothelial cells, often intravascular. **Masson's pseudoangiosarcoma** usually represents an organized thrombus that shows papillary proliferations of endothelial cells, as well as irregular vascular channels, and can also resemble angiosarcoma.

Malignant Vascular Tumors

Kaposi's Sarcoma (KS)

There are three types:

1. a classic form seen in elderly patients, particularly of Mediterranean ancestry
2. an aggressive form seen in Central Africa, affecting both young and middle-aged subjects, possibly related to Epstein-Barr virus
3. a form seen in immunocompromised patients, first described in renal transplant recipients, and later in HIV (AIDS) infected individuals.

21

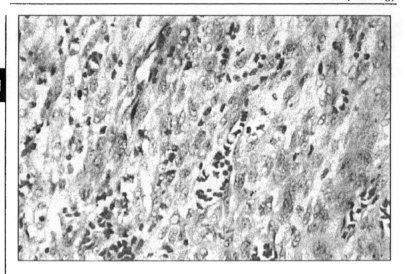

Fig 21.10. **Kaposi's sarcoma, classic**: Proliferation of spindle cells with vesicular nuclei forming slit-like spaces with extravasation of red blood cells.

Clinical

Three stages, macular, plaque, and nodular, are seen. Involvement of regional lymph nodes, mucosas, and internal organs are possible. In the classic form, purpuric macules are seen, particularly on the feet and legs. The macules evolve into plaques and nodules. There is a slow progression, and the tumor may eventually involve lymph nodes, as well as internal organs. In AIDS-related Kaposi's sarcoma, the presentation is more variable, involving skin or mucosas (commonly the oral mucosa). It usually presents as purpuric macules, resembling ecchymosis, although tumor nodules can be seen from presentation.

Histology

The classic, diagnostic histologic picture of KS is the plaque or tumor stage characterized by a proliferation of spindle cells that run roughly parallel to each other, although they do not display any characteristic architectural arrangement. Among the tumor cells multiple extravasated RBCs are seen, as well as slit-like vascular spaces. The individual tumor cells show atypical nuclear features, and mitosis are often seen (Fig. 21.10).

Early, macular KS is more subtle. There is a proliferation of endothelial cells and small blood vessels throughout the dermis. A key finding is the presence of hemosiderin deposits at all levels of the dermis. A characteristic finding is the presence of eosinophilic hyalin round bodies in the cytoplasm of the tumor cells that can be highlighted by staining with PAS. These bodies may represent phagocytized RBCs. The tumor infiltrates the periadnexal dermis, seen between eccrine coils in a characteristic fashion. In addition to the spindle cell component, dilated vascular spaces and/or a capillary-like proliferation resembling granulation tissue can also be seen. In the upper dermis, jagged vascular spaces may be seen,

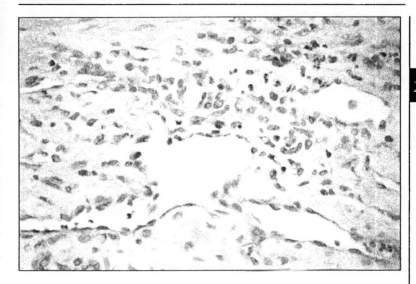

21

Fig 21.11. **Early Kaposi's sarcoma**: Jagged vascular spaces that are focally infiltrating the dermis, prominent endothelial cells and a lymphoplasmacytic infiltrate.

and they may be surrounded by an infiltrate of lymphocytes and plasma cells. Involvement of pre-existing vessel walls is also seen in early KS (Fig. 21.11).

Differential Diagnosis

Old hematomas or hemorrhage in the skin will also show hemosiderin deposits throughout the dermis. Acroangiodermatitis shows lobular proliferation of capillaries, rather than diffuse infiltration as seen in KS. Occasionally scars show granulation tissue and hemosiderin deposits that may resemble Kaposi's sarcoma. In those cases, however, the dermis shows fibrosis and young fibroblasts, which are absent in the case of KS.

Angiosarcoma

Clinical

Angiosarcomas occur in two different settings: as a result of lymphatic or vascular obstruction, as is the case in postmastectomy lymphangiectasia (Stewart-Treves syndrome), and as a spontaneous tumor in elderly patients on the head and neck. The former setting is clinically manifested by lymphedema and subsequent formation of tumor nodules. In the case of older patients it may manifest only as a vascular macular proliferation that enlarges and can become a plaque.

Histology

In the macular or plaque tumors of elderly patients, the histologic findings can be quite subtle and are characterized by the presence of anastomosing vascular channels in the papillary dermis, which are usually lined by plump, slightly atypical endothelial cells. The vascular channels are often also infiltrating the reticular der-

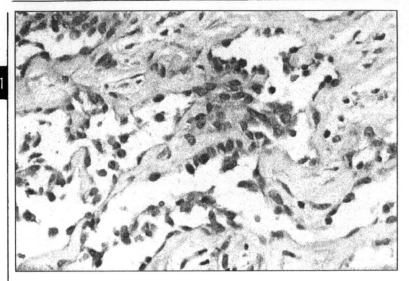

Fig 21.12. **Angiosarcoma**: Anastomosing vascular channels lined by hyperchromatic large atypical endothelial cells, infiltrating the dermis.

mis, as well as the subcutaneous tissue, but their presence may be overlooked since they may be confused with artifactual spaces separating collagen bundles. In the case of tumor nodules, the histologic appearance is that of a malignant, often solid tumor in which vascular spaces are lined by highly atypical endothelial cells (Fig. 21.12). There are also solid areas with mainly a proliferation of atypical endothelial cells. In the solid areas, intracytoplasmic lumina are seen in the neoplastic endothelial cells. The tumor cells may show vacuolated round or oval nuclei and clear cytoplasm.

Stains

With immunoperoxidase stains, tumor cells are positive for endothelial markers, factor VIII-RA, CD-34, and CD-31 and they are also positive for Vimentin.

Hemangiopericytoma

Clinical

Hemangiopericytomas are usually soft tissue tumors that can involve the skin secondarily. They are seen on the extremities and on the trunk.

Histology

There are several tumors made of pericytes, the sustentacular cells outside of the basal membrane of capillaries, with support and possible contraction properties. Hemangiopericytomas are very cellular tumors with cells exhibiting homogeneous nuclei and spindle cell cytoplasm. They are arranged in a perivascular distribution, surrounding multiple ramifying (staghorn) vascular spaces, which are lined by normal endothelial cells.

Stains

Reticulin stain will show tumor cells separated from the endothelium of the vascular spaces. The behavior is unpredictable. The presence of mitosis, particularly atypical mitosis, as well as some degree of necrosis may indicate malignancy.

Section 58. Neural Tumors

Neural tumors are common neoplasms of the skin. Usually an incidental finding, they may portend the presence of neurofibromatosis. Most neural tumors are benign, and classically they have been separated into three categories: neuroma, neurilemmoma or Schwannoma, and neurofibroma. Neuromas are hamartomas or hyperplasia of small nerves, which usually form as reactions to trauma. Schwannomas are benign tumors composed of Schwann (sustentacular cells) and are true encapsulated tumors. Neurofibromas are mixed tumors composed of Schwann cells and interstitial and perineural fibrous tissue (nerve sheaths), and usually they contain trapped axons.

Neuromas

Clinical

Morton's neuromas occur in the third metatarso-phalangeal space, usually due to trauma. Amputation neuromas ensue after severance of nerve trunks or bundles. Multiple neuromas presenting as small nodules have been described on lips and oral mucosa, usually associated with multiple endocrine tumors (MEA 2b), i.e., mucosal/neuroma syndrome. Pallisading and encapsulated neuroma (Reed) occurs mainly on the face.

Histology

Neuromas exhibit normal appearing or hyperplastic nerve bundles surrounded by fibrous tissue (scar). Hyperplasia of the perineural fibrous tissue is common. The size of the nerve bundles is variable and may be quite small. In the case of encapsulated neuroma there is one tumor nodule with peripheral condensation of the collagen. Nerve fascicles with rough nuclear pallisading are seen (Fig. 21.13).

Neurilemmoma (Schwannoma)

Clinical

Benign tumors manifest in the skin as small papules or nodules. They can be deep in the subcutaneous tissue. They are the second most common tumors in neurofibromatosis, after neurofibromas. If large enough, they can be seen grossly as a lateral expansion of a nerve.

Fig 21.13. **Neuroma**: Proliferation of nerves and nerve bundles surrounded by fi-brous tissue.

Histology

Well-delimited, encapsulated tumors that show classically two histologic pat-terns, randomly distributed throughout the tumor. The first one is a loose, mixoid-like component characterized by spindle cells mixed with mucinous , loose connective tissue background. This has been termed Anthoni B. The second pattern is characterized by more fibrous and densely cellular areas, in which spindle cells pre-dominate. It is called Anthoni A, and in these areas the nuclei of the cells are often arranged in orderly rows, or palisades. Two such rows opposed to each other, with an acellular background substance in between cytoplasm, form what is called Verocay's bodies. They are diagnostic of neurilemmoma (Fig. 21.14).

Plexiform neurilemmomas (Schwannoma)resemble plexiform neurofibromas by the presence of nerve bundles. However they also show Verocay's bodies, are not associated with neurofibromatosis, and do not become malignant.

Note: The presence of Verocay bodies in a plexiform neurofibroma suggests that the tumor is indeed a Schwannoma.

Stains

S-100 and CD-57 stains will be positive in neurilemmomas.

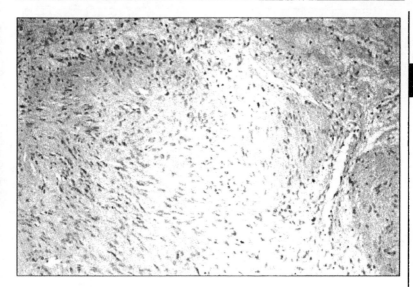

21

Fig 21.14. **Neurilemmoma**: Antoni A areas of fibrous, spindle cells with formation of Verocay bodies are surrounded by loose, myxoid Antoni B areas.

Neurofibromas

Clinical

Most commonly present as solitary papules without any predilection for location. Clinically, they resemble melanocytic nevi. On cut section they show a white, soft, glistening surface. When associated with neurofibromatosis, hundreds, or even thousands, of neurofibromas may be apparent as papules or nodules of different sizes. In the involved areas the skin may appear folded and baggy, and also hyperpigmented (molluscum pendulum). Neurofibromatosis is also associated with other tumors, including neurilemmomas, with cafe-au-lait spots, axillary freckling, and visual and auditory tumor involvement. Deep seated tumors are common, and malignant transformation is a possibility.

Histology

Most neurofibromas are solitary, incidental findings. They are small and are characterized by spindle cells that appear wavy, and do not display any specific architectural arrangement. The tumors are generally located throughout the dermis. They do not displace cutaneous adnexa, but surround them, and are quite well circumscribed. The nuclei are relatively small, homogeneous, and spindle, and the cytoplasmic margins are not apparent (Fig. 21.15). **Neurothekomas**, or nerve sheath myxomas, are lobulated tumors characterized by small spindle cells admixed in a mucinous stroma. **Pallisading and encapsulated neuroma of Reed** is usually more densely cellular, resembling a neuroma. The nuclei are often arranged in rows or palisades, and there is a thin fibrous tissue capsule surrounding the tumor. **Plexi-**

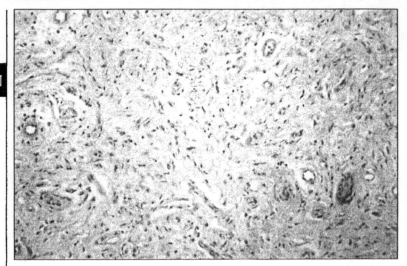

Fig 21.15. **Neurofibroma**: A proliferation of spindle cells, many of them with wavy nuclei, admixed with fibrous tissue and small blood vessels.

form neurofibromas, which can reach a large size show on gross examination a "bag of worms" appearance. Histologically they exhibit enlarged nerve bundles (intraneural neurofibroma). Plexiform neurofibromas may be diagnostic of neurofibromatosis.

Stains

Neurofibromas are positive for S-100 and Leu-7. CD-57 Bodian stain will show nerve fibers among the tumor cells in neurofibroma, but not in neurilemmoma.

Differential Diagnosis

Neurofibromas can be differentiated from dermatofibromas by lack of epidermal hyperplasia, lack of storiform pattern and of giant cells, by being well circumscribed and by the presence of cutaneous adnexa within the tumor.

Granular Cell Tumors

Clinical

Generally present as small papules and nodules; rarely as verrucous plaques. The tumors are frequently seen on the tongue, but they have been described as occurring everywhere on the skin and on most organs. They can be solitary or multiple.

Histology

There is a proliferation of large cells with a small nucleus and a granular distended cytoplasm. The cells infiltrate throughout the dermis and often involve in the walls of blood vessels, or the arrector pilaris muscle (Fig. 21.16). Granular cell tumors are considered to be of neural origin as opposed to muscle origin, as was first

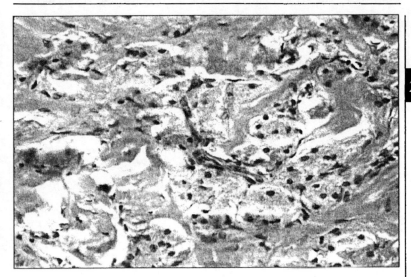

Fig 21.16. **Granular cell tumor**: A collection of cells with large, granular cytoplasm and small, regular hyperchromatic nuclei are seen with PAS stain.

thought. Actually the exact nature of the tumor cells is not well understood, and they may represent a degenerative process, with more than one type of cell origin.

Stains
The granules stain positive with PAS and S-100. Under the electron microscope they are phagolysosomes.

Other Tumors
Bednar tumor represents a pigmented dermatofibrosarcoma protuberans (DFSP). The tumor exhibits the same cellular configuration as ordinary dermatofibrosarcoma protuberans, except that some of the cells are loaded with melanin. The pigment stains positive with Fontana and negative with Perls. An intriguing theory is that DFSPs are of neural origin.

Nasal gliomas are really hamartomas, resulting from the trapping of glial tissue in the skin of the nose during fetal life. The tumor grows at the same rate as the body, and it is manifested by a nodule somewhere along the midline on the root or the tip of the nose. Histologically, glial tissue is seen replacing the dermis and subcutaneous tissue, manifested by fibrillary material with discrete nuclei. Immunoperoxidase stain is positive for GFAP. **Cutaneous meningioma** usually represents a herniation of meninges that remained in the skin after bone closure during fetal life. If there is a persistent bone defect, then meningoceles and meningomyeloceles ensue. Histologically cutaneous meningiomas show histologically meningeal tissue, often with formations of meningeal eddies, and presence of psammoma bodies. They stain with EMA, while S-100 and cytokeratin expression is variable.

Neurofibrosarcoma

Clinical

It is seen mainly in the setting of neurofibromatosis. It may manifest by a rapid growth of a pre-existing neurofibroma.

Histology

It is characterized by a cellular neurofibroma, in which mitoses are seen (beware of any mitosis in a neurofibroma), and nuclear atypia is present. Formation of terminal body-like neural structures may be seen. Necrosis may be present. Although the term "malignant Schwannoma" has been used for this tumor, it represents a malignant transformation of a neurofibroma.

CHAPTER 22

Tumors of Lymphoid Tissue, Metastatic Disease

Ramón L. Sánchez and Sharon S. Raimer

Section 59. Lymphoid Infiltrates and Tumors

We include in this chapter lymphoid and hematopoietic tumors that have primary or secondary involvement of the skin as well as benign lymphocytic infiltrates that, because of the clinical and histologic findings, should be differentiated from malignant lymphoid tumors.

Benign Lymphocytic Infiltrates and Pseudolymphomas

We have grouped under this category several entities which are characterized by large lymphoid aggregates in the dermis without cytologic atypia. Some of these infiltrates, namely polymorphous light eruption, Jessner's lymphocytic infiltrate, and persistent insect bites, have been discussed elsewhere (Chapter 8) and are characterized by patchy lymphoid aggregates, that are perivascular and periadnexal, superficial and deep. Occasionally the infiltrate is so prominent that it involves a large proportion of, or even all of, the dermis. There is often a nodular architectural arrangement, and formation of lymphoid follicles or germinal centers can be seen. In addition to mature lymphocytes, other cells frequently present are plasma cells, eosinophils, larger macrophage-histiocytic cells, as well as prominent endothelial cells. The keys to the diagnosis are the lack of atypical cytologic features, as well as the polymorphism of the infiltrate. We prefer to use the term lymphoid hyperplasia since pseudolymphoma may be misleading. In general, benign lymphoid infiltrates of the skin are top heavy as opposed to malignant lymphomas that are bottom heavy.

Reactive Lymphoid Hyperplasia

Clinical

Characterized by papules, pink to whitish, particularly on the face. They may remain the same size after a period of growth. Clinically they should be differentiated from other infiltrative processes including sarcoidosis, adnexal tumors and true lymphomas.

Histology

There is a marked lymphoid infiltrate that involves most of the dermis and has a nodular arrangement with formation of lymphoid follicles and an attempt to form germinal centers. In the latter areas there is a mixture of cells including macrophages, lymphoid cells (in different stages of maturation), plasma cells, and occasional

Dermatopathology, edited by Ramón L. Sánchez and Sharon S. Raimer. ©2001 Landes Bioscience.

22

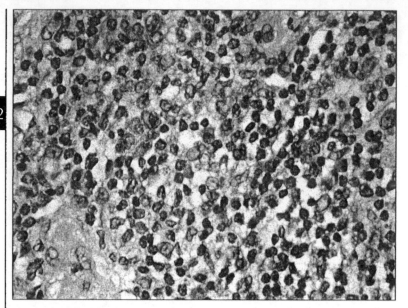

Fig 22.1 . **Lymphoid hyperplasia**: Prominent lymphohistiocytic infiltrate with plasma cells and eosinophils. Attempt to form lymphoid follicles is likely seen.

eosinophils. Tangible body macrophages that have phagocytosed nuclear fragments, can be seen (Fig. 22.1). Histologically the process should be differentiated from well-differentiated lymphomas of B-cell origin, such as the marginal zone lymphoma or lymphoma of skin associated lymphoid tissue (SALT lymphoma). The differential diagnosis with these lymphomas may be difficult to make and may require molecular diagnostic techniques. Occasionally reactive lymphoid infiltrates, particularly when the result of some immunologic stimulus such as persistent insect bite, may show markedly atypical cytologic features, which can be difficult to differentiate from malignant lymphomas. Kikuchi's disease, a benign, probably viral-induced process, also shows an atypical lymphoid infiltrate in the skin.

Sinus Histiocytosis with Massive Lymphadenopathy (SHML)

Clinical
This entity described originally by Rosai and Dorfman in lymph nodes and other organs can also be seen in the skin. Those cases are clinically characterized by yellowish or pink papules, particularly on the trunk or extremities.

Histology
SHML recapitulates in the skin the finding seen in the lymph nodes. There is an extensive and marked infiltrate of lymphocytes, plasma cells, and eosinophils. The characteristic cells are macrophages that display large clear or foamy cytoplasm. The

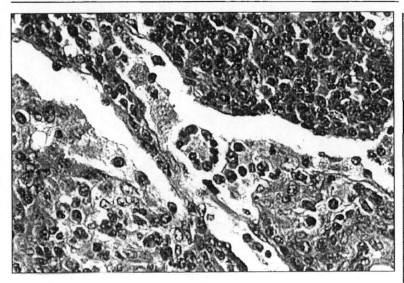

22

Fig 22.2. **Sinus histiocytosis with massive lymphadenopathy**: Macrophages admixed with lymphocytes and giant cells, some of which have internalized inflammatory cells.

macrophages are arranged in a pattern that is reminiscent of the sinus in a lymph node. There is frequent phagocytosis or emperipolesis of lymphocytes, PMNs and RBCs by the macrophages. The lymphocytes line the areas with macrophages as the trabeculae in the lymph nodes (Fig. 22.2).

Stains
The macrophages in SHML stain positive for S-100.

Lymphomatoid Papulosis

Clinical
Papules, particularly on the extremities and on the trunk that reach a size of one centimeter or slightly larger, develop a necrotic center, and involute in a period of six weeks. New papules develop while the old ones resolve.

Histology
Two types, A and B. Type A is characterized by a lymphoid infiltrate among which large atypical cells are seen. The infiltrate is usually distributed in patches in the dermis and it may be perivascular. Invasion of the epidermis is possible (Fig. 22.3).

Type B lymphomatoid papulosis is characterized by a histology similar to that seen in mycosis fungoides. There are moderately atypical lymphocytes, possibly with convoluted nuclei, which display epidermotropism. Pautrier microabscesses are seen.

22

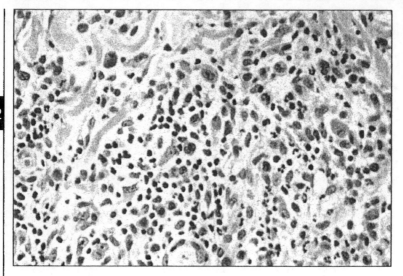

Fig 22.3. **Lymphomatoid papulosis**: Predominantly a lymphocytic infiltrate with large, atypical cells which stain positive with CD30 (Ki-1) antigen.

Stains
The large atypical cells on type A lymphomatoid papulosis often stain positive for CD-30 (Ki-1).

Differential Diagnosis
The histologic features in lymphomatoid papulosis should be differentiated from large cell anaplastic lymphoma, as well as from mycosis fungoides. Clinical and pathologic correlation is extremely important to establish the correct diagnosis. Perhaps as many as 10% of patients with lymphomatoid papulosis will eventually develop an overt lymphoma.

Lymphomas
Primary lymphomas of the skin are both of B- and T-cell origin. Since the resident lymphoid tissue of the skin, which is normally located in the papillary and periadnexal dermis, is T-cell, it follows that more primary lymphomas of the skin are of T-cell origin than of B-cell origin.

Lymphomas: B-Cell Type Large Cell Immunoblastic (Large B-Cell Lymphoma)

Clinical
Characterized clinically by individual nodules or plaques, particularly on the extremities in middle-aged or older patients. Intermediate prognosis.

Histology

There is a monomorphic infiltrate with large lymphocytes that display a round to oval nucleus with one to two prominent purplish nucleoli. A rim of amphophilic cytoplasm is often seen. The infiltrate is quite well delineated and involves full thickness of the dermis, extending to epidermis and subcutaneous tissue (Fig. 22.4).

Stains

Immunoperoxidase shows a monoclonal population of B-cells. Light chains gamma and kappa stains may be helpful.

22

Well-Differentiated Lymphocytic Lymphoma (Follicular Center Cell Lymphoma)

Clinical

There are one or more red nodules in the skin of middle aged to elderly patients. Often on the head and trunk. Good prognosis.

Histology

A rather monomorphic lymphocytic infiltrate is seen throughout the dermis with follicular or nodular formation. The tumor cells resemble normal mature lymphocytes. The diagnosis is given by the monomorphism, the prominence of the infiltrate, and the replacement of normal structures by the tumor cells.

Stains

Immunoperoxidase stains may help to demonstrate a monoclonal B-cell infiltrate.

Lymphomas: T-Cell Type (CTCL)

Primary T-cell lymphomas of the skin have been designated under the umbrella designation of "cutaneous T-cell lymphoma" (CTCL). While this is an accurate term, it does not indicate specific differences among T-cell lymphomas. The main entity in this section, both by frequency of appearance and by the typical clinical presentation, is mycosis fungoides. Traditionally, a premycosis stage has been described under the rubric of parapsoriasis, with multiple specific designations according to its clinical characteristics. Other T-cell lymphomas of the skin-nonmycosis fungoides-have been recently recognized and given individual identities.

Large Plaque Parapsoriasis

Clinical

Large plaque parapsoriasis is usually considered to be a premalignant condition. Most cases show atrophic, hyperkeratotic poikilodermatous plaques. **Parapsoriasis variegata** can present as nonatrophic plaques. The two most common types of small plaque parapsoriasis are **digitate parapsoriasis** and **pityriasis lichenoidis chronica** (discussed elsewhere in this text).

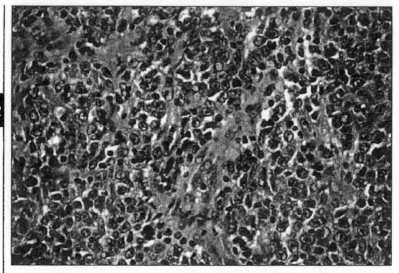

Fig 22.4. **Large cell immunoblastic B-cell lymphoma**: Sheets of lymphoid cells with large vesicular nucleus with prominent nucleolus and scanty cytoplasm.

Histology

Large plaque parapsoriasis exhibits the changes of poikilodermatitis, with hyperkeratosis, atrophy of the epidermis, hydropic degeneration of the basal layer, and a mild lymphocytic infiltrate around the superficial capillary plexus, which extends into the papillary dermis and may show focal epidermotropism (Fig. 22.5). Some of the lymphocytes may appear convoluted or cerebriform. Other changes seen in the papillary dermis are pigment incontinence and dilatation of capillaries. **Parapsoriasis variegata** may show normal to acanthotic epidermis, and a band-like infiltrate in the upper dermis. Some of the lymphocytes may show exocytosis or epidermotropism. Occasionally parapsoriasis variegata shows large deposits of hemosiderin among the infiltrate, a manifestation of previous hemorrhage.

Mycosis Fungoides (MF)/Sezari's Syndrome

Clinical

Three stages are identified in the evolution of mycosis fungoides: patch stage, plaque stage and tumor stage. Generally patients stay in patch or plaque stage for many years, sometimes even before the disease is properly diagnosed. While in these first two stages patients respond quite well to topical treatments. Patients in the third stage, the nodular or tumoral stage, usually have a more rapid evolution, and the prognosis is much worse.

Hypopigmented MF refers to an early presentation of the disease, characterized by hypopigmented macules with only mild scale.

Sezary's syndrome presents clinically with erythroderma or at least with redness and congestion, particularly of the face. Occasionally patients with mycosis fungoides can also present with erythroderma. Some patient in the tumor stage present with **leonine facies**. **Actinic reticuloid** may likewise present with leonine facies. The tumor stage of MF and Sezary's syndrome are advanced stages of mycosis fungoides. Patients with clinical lymphadenopathy with or without histologic involvement of the lymph nodes by mycosis fungoides are also placed in a high, advanced stage. **Mycosis fungoides d'emblee** debuts as nodular lymphoma which tends to progress rapidly. Variations of the classical appearance of mycosis fungoides include a **granulomatous type** which should be differentiated from "granulomatous slack skin syndrome" (a type of T-cell lymphoma of the skin) and **psoriasiform mycosis fungoides**.

22

Histology

The infiltrate of mycosis fungoides varies from fairly sparse in the patch stage to very prominent in the tumor stage. Likewise the histologic features of the tumor cells evolve from slightly atypical, although cells are convoluted or cerebriform in the patch stage, to large atypical blast-like cells in the tumor stage, or a combination of both. In the earlier stages the following histologic features help to establish the diagnosis of mycosis fungoides:

1. band-like lymphocytic infiltrate, which is superficial, with deeper patches of lymphocytes in the dermis. Eosinophils are usually seen among the lymphocytes.
2. Epidermotropism of lymphocytes into the epidermis. Epidermotropism is distinguished from exocytosis by the absence of spongiosis in the former. The epidermotropic cells are surrounded by a clear halo, but the adjacent keratinocytes do not show edema.
3. A very early change is a linear sprinkling of lymphocytes along or just above the basal layer of the epidermis, each of them surrounded by a clear halo. In this particular instance there is often pigment incontinence or loss of pigmentation of the epidermis, which is translated clinically by hypopigmented flat patches seen in large plaque parapsoriasis or in hypopigmented mycosis fungoides (Fig. 22.6).
4. The presence of **Pautrier microabscesses** in the epidermis. The Pautrier microabscesses should be differentiated from spongiotic vesicles seen in the context of eczema. They appear as a collection of atypical lymphocytes in the stratum of Malpighii (Fig. 22.7).

Note: The presence of an atypical lymphocytic infiltrate with focal epidermotropism may not be diagnostic. In those cases it is better to make a diagnosis of atypical lymphocytic infiltrate suspicious, but not diagnostic of mycosis fungoides. A rebiopsy may later establish the definitive diagnosis.

In tumor stage mycosis fungoides, the infiltrate has increased considerably and is now involving most of the dermis. Many of the cells of the infiltrate have undergone a blastic or more primitive transformation, showing large nuclei with small cytoplasmic outline and often one or two nucleoli (Fig. 22.8).

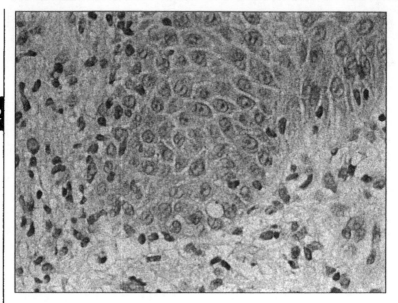

Fig 22.5. **Hypopigmented mycosis fungoides**: A sprinkle of lymphocytes are invading the epidermal basal layer with halo formation and loss of pigment.

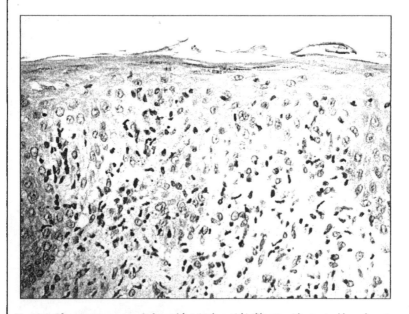

Fig 22.6. **Plaque stage mycosis fungoides**: Lichenoid infiltrate with atypical lymphocytes that are invading the epidermis (epidermotropism). Early Pautrier microabscess.

22

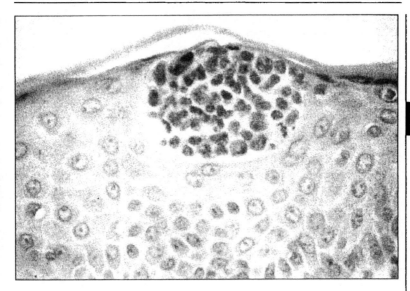

Fig 22.7. **Pautrier microabscess:** A collection of atypical lymphocytes in the epidermis is diagnostic for mycosis fungoides.

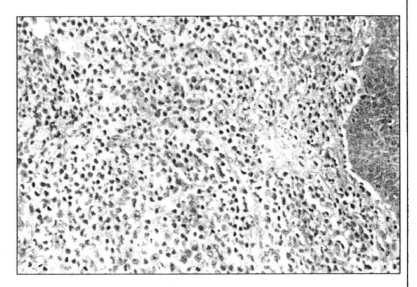

Fig 22.8. **Mycosis fungoides, tumor:** The dermis is completely infiltrated by atypical lymphocytes, many of which show large, vacuolated nuclei.

Sezary's syndrome may show only scattered lymphocytes with atypical features similar to those seen in the patch stage of mycosis fungoides. The diagnosis of Sezary's syndrome is usually made by examination of the peripheral blood.

Stains

Immunoperoxidase stains are helpful to establish the T-cell nature of the infiltrate. Immunoperoxidase will also clearly delineate the lymphocytes involving the epidermis. With fresh tissue and monoclonal antibodies, a majority of the cells of the infiltrate will be CD-3 (Pan-T) positive cells with absence of CD-7 or CD-5 surface antigens. Similar results are obtained by means of flow cytometry with peripheral blood in Sezary's syndrome.

Note: The absence of CD-5 and/or CD-7 among a majority of cells of the infiltrate, while being CD-3 positive, is diagnostic for mycosis fungoides or Sezary's syndrome. T-cell rearrangement tests are also presently used.

Adult T-Cell Leukemia/Lymphoma

Clinical

Associated with HTLV-1 virus, endemic in some areas of Japan and in the West Indies, very rare in the United States or Europe. Patients present with multiple papules distributed throughout the body, including head and neck. Changes in peripheral blood include hypercalcemia and high alkaline phosphatase.

Histology

There are multiple intraepidermal abscesses resembling Pautrier microabscesses made by neoplastic atypical lymphocytes, which are also infiltrating the dermis.

Anaplastic Large-Cell Lymphoma, Ki-1 Positive (ALCL)

Clinical

There is a spectrum of clinical lesions seen, from one or a few papules that may develop ulcerated centers and resemble lymphomatoid papulosis to multiple papules and nodules. The lesions do not self-resolve as in the case of lymphomatoid papulosis.

Histology

The dermis is infiltrated by large cells with atypical, vacuolated nuclei and abundant clear cytoplasms (Fig. 22.9).

Stains

Most anaplastic large cell lymphomas show a T-cell phenotype, as demonstrated with immunoperoxidase. Occasionally they may be B-cell tumors. All of them are Ki-1 (CD-30) positive.

Note: Anaplastic large cell lymphoma was originally described as occurring in lymph nodes, and later described as occurring in the skin. It may be at the end of the spectrum opposite to lymphomatoid papulosis, although it is not clear that cases of lymphomatoid papulosis may eventually become ALCL. CD-30 is positive in the cells of ALCL, as well as in the large atypical cells of lymphomatoid papulosis, and

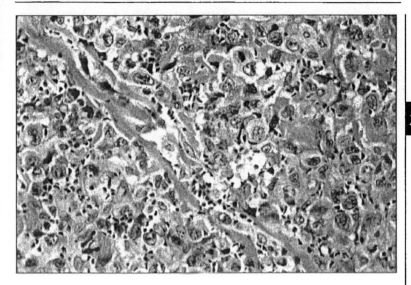

Fig 22.9. **Anaplastic large cell lymphoma**: The tumor cells are very large with abundant clear cytoplasm and very irregular nuclei. Tumor is Ki-1 (CD30) positive.

in some Reed-Sternberg cells in Hodgkin's disease. Ki-1 (CD-30) is a marker of cell activation, both B- and T-cells may express this marker.

Lymphoblastic T-cell Lymphoma

Clinical
Single or multiple nodules and plaques can be seen throughout the body. The clinical appearance can resemble chronic dermatitis or an infectious process.

Histology
There is an atypical lymphoid infiltrate in the dermis which may involve and replace the full dermis, or that may be patchy and arranged around hair follicles and other cutaneous adnexae. It sometimes demonstrates a perineural arrangement. The subcutaneous fat is involved in most instances. Epidermotropism may be inconspicuous (Fig. 22.10).

Note: The infiltrate in lymphoblastic T-cell lymphoma may not show marked cytologic atypia. Often a high degree of suspicion is necessary to make the diagnosis. Features that help in the diagnosis are the clear, vacuolated nuclei that can be round, oval or elongated, and clear cytoplasm seen in most cells. Some eosinophils and mature lymphocytes are also admixed in the infiltrate.

Stains
Immunoperoxidase stains are helpful to demonstrate the T-cell nature of the neoplastic cells. Molecular techniques demonstrate T-cell rearrangement and are diagnostic.

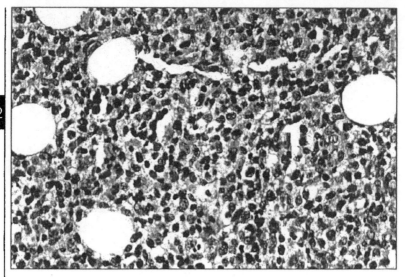

Fig 22.10. **Lymphoblastic T-cell lymphoma**: T-cell lymphoma with atypical lymphocytes which is invading the subcutaneous tissue and lower dermis.

Angiocentric Lymphomas, Lymphomatoid Granulomatosis

Clinical
Angiocentric lymphomas can involve the skin, presenting as nodules or plaques that can be ulcerated. Some of these patients also have pulmonary involvement, which may be the first manifestation of the disease. Renal involvement may also be present. Some of these patients carry a diagnosis of **lymphomatoid granulomatosis**.

Histology
The skin biopsy shows an atypical infiltrate involving the subcutaneous tissue as well as the dermis. Necrotizing granulomas are usually recognized. The infiltrate involves the wall of mid-sized veins and/or arteries where granulomatous inflammation can also be seen.

Subcutaneous T-cell Lymphomas

Clinical
This entity, presents clinically with nodules or ulcers, often in the extremities that appear vascular or infectious in nature.

Histology
The infiltrate is localized in the subcutaneous tissue and secondarily in the dermis. The cytologic features are variable and in some cases the cells are large and only slightly atypical, resembling macrophages. Hemophagocytosis can be seen, which can lead to a misdiagnosis of histiocytic cytophagic panniculitis. Other cases show

tumor cells that are highly atypical with large vacuolated nuclei and abundant mitosis, many of them atypical (Fig. 22.11).

Midline Granuloma Syndrome

Clinical
Centrophacial lymphomas can present with a variable degree of involvement of the respiratory airways, which can lead to destruction of these structures.

Histology
These lymphomas are both of B-cell or T-cell origin, and they are often characterized by large atypical lymphoid cells or immunoblasts. Tumor cells may be mixed with normal lymphocytes, and the necrosis of tissue may produce a considerable amount of secondary inflammation, obscuring the diagnostic features. Granuloma formation can be seen, particularly in the setting of lymphomatoid granulomatosis. Some of these lymphomas have been associated with E-B virus infections.

Malignant Angioendotheliomatosis

Clinical
Characterized by erythematous eruption and small papules or plaques. Central nervous system involvement is common.

Histology
The blood vessels in the dermis are dilated and filled with atypical lymphocytes. This entity represents an intravascular invasion by lymphoma cells, usually of B-cell origin. It should be differentiated from reactive angioendotheliomatosis, a benign process with similar histologic features, in which the cells filling the blood vessel lumina are endothelial cells.

Stains
Immunoperoxidase stains can help in the diagnosis by showing lymphocyte markers in the case of lymphoma, while they are positive for endothelial markers in reactive angioendotheliomatosis.

Leukemias

Clinical
Acute lymphoblastic leukemia, seen mainly in children, can present clinically as multiple nodules or plaques with a wide distribution.

Histology
There is a dense lymphoid infiltrate which is effacing the normal cutaneous architecture resembling any of the other small blue cell tumors in the skin.

The nature of the cells is confirmed by immunoperoxidase which will show positive lymphoid markers. In most instances it is a T-cell lymphoblastic leukemia. It cannot be differentiated from lymphoblastic lymphoma, although the clinical setting and the peripheral blood count would suggest the diagnosis. The diagnosis of leukemia is confirmed by a bone marrow biopsy.

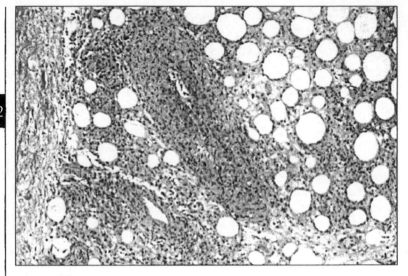

Fig 22.11. **Subcutaneous T-cell lymphoma**: This tumor may resemble a panniculitis, particularly if the cytologic features are not markedly atypical.

Acute Myelogenous Leukemia

Clinical
Acute myelogenous leukemia can manifest in the skin with one or more nodules that have a bluish discoloration.

Histology
The infiltrate is more diffuse than that seen in a lymphoma, but if the cells are very primitive they cannot be easily distinguished from neoplastic lymphocytes. Immunoperoxidase stains for paraffin-embedded tissue may show positivity for CD-45 (leukocyte common antigen) and for T-cell markers, which suggest the diagnosis of lymphoma. However, the neoplastic cells will also be positive for CD-68, lysozyme, and myeloperoxidase, (macrophage and granulocytic markers) which establishes the diagnosis of myelogenous leukemia rather than that of lymphoma. There is also positivity for CD43.

Note: When a malignant infiltrate is seen in the skin, the inclusion of macrophage markers in the panel of immunoperoxidase stains will allow the differential diagnosis of leukemia versus lymphoma.

Chronic Lymphocytic Leukemia
Chronic lymphocytic leukemia, when it involves the skin, cannot be differentiated histologically from well-differentiated lymphocytic lymphoma. It is usually seen in older patients, which may also present with lymphadenopathy. Histologically, a marked infiltrate of mature lymphocytes replaces the dermis.

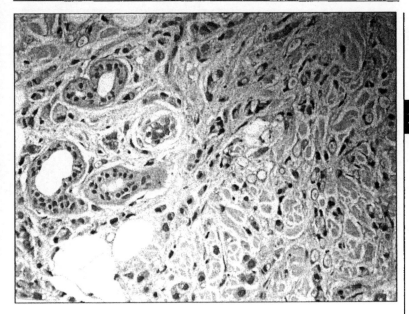

22

Fig 22.12. **Metastatic adenocarcinoma to the skin**: Many of the tumor cells show accumulation of mucin in their cytoplasm with signet ring cell appearance.

Section 60. Metastatic Disease

Metastatic tumors in the skin may originate from any source; however, there are several tumors that metastasize more frequently to the skin. Among them are metastatic melanoma, carcinoma of the lung, from the gastrointestinal tract, kidney, and prostate for men, and breast and lung for women. Histologically the diagnosis of metastatic malignancy is made by the presence of malignant tumor cells or nests that are infiltrating the dermis in a haphazardly fashion without any architectural arrangement. The metastatic tumor often destroys cutaneous adnexa and can be seen dissecting between collagen bundles (Indian file) or inside of blood vessels or dermal lymphatics (commonly seen in breast carcinoma). Metastatic melanoma is usually characterized by nests of melanoma cells without a junctional component. (Be aware of metastatic melanoma, when is connected to the epidermis, resembling a primary tumor). Metastatic carcinomas, particularly when seen on the head, neck, or upper trunk, often come from the larynx, pharynx, esophagus, or lungs. The degree of differentiation is variable. Metastatic carcinomas from the lung, however, may appear as poorly differentiated carcinomas, such as the large anaplastic type. Small cell carcinomas of the lung have to be differentiated from Merkel cell tumors, and display similar nuclear and cytologic characteristics. Renal cell carcinomas, although rare, show clear cells that stain positive with vimentin and keratin.

Table 22.1. Most common metastatic tumors to the skin

Carcinoma of the lung
Carcinoma of the breast
Melanoma
Carcinoma of oral cavity, pharynx, larynx
Carcinoma of gastrointestinal tract
Carcinoma of the ovary
Renal cell carcinomas
Carcinoma of prostate

22

CHAPTER 23

Special Stains and Techniques

San-Hwan Chen

Section 61. Techniques

A. Histochemical Techniques

Connective Tissue Stains

	Stains	Uses	Results
1.	Hematoxylin-eosin	Routine	Nuclei, calcium: blue Collagen, Muscles, Nerves: red
2.	Masson trichrome	Collagen Spindle cell tumors (Leiomyoma vs dermatofibroma)	Collagen: blue or green Muscles, Nerves, Nuclei: red
3.	Verhoeff-van Gieson*	Elastic fibers Reactive perforating collagenosis	Elastic fiber: black Collagen: red Nuclei, Muscles, Nerves: yellow
4.	Acid orcein*	Elastic fibers	Dark brown
5.	Gomori's aldehyde fuchsin*	Elastic fibers	
6.	Movat (pentachrome)	*All three: Pseudoxan- thoma elasticum, connective tissue nevus, erythema ab igne, elastosis perforans serpiginosa, annular elastolytic granuloma, cutis laxa, polyarteritis nodosa vs migratory thrombophlebitis, nodular vasculitis	Nuclei, Muscle: red Collagen: yellow Elastic fibers: black AMP: green

Mineral Stains

	Stains	Uses	Results
7.	Von Kossa*	Calcium	Black
8.	Alizarin red S*	Calcium *Both: calcinosis cutis, pseudoxanthoma elasticum, subepidermal calcified nodule, pilomatricoma, malakoplakia	Orange-red

9.	Perls potassium ferrocyanide	Hemochromatosis, dermatofibroma, pigmented purpuric dermatoses, drug-induced pigmentation, stasis dermatitis, Monsel's reaction	Hemosiderin: Blue

Pigment Stains

10.	Fontana-Masson	Melanin (argentaffin) *Melanoma, metastatic cutaneous carcinoid, drug-induced pigmentation	Melanin and Axgentaffin Granules Black; Nuclei: red
11.	Dopa reaction (in unfixed tissue)	Tyrosinase in melanocytes *vitiligo	Brown-black
12.	Silver nitrate (Grimelius)	Melanin, reticular fibers (argyrophilic) Some carcinoids, pheochromocytomas	Black

Carbohydrate Stains

13.	Periodic acid-Schiff reaction (PAS)	Glycogen (diastase-labile) *Bowen's disease, clear cell acanthoma, trichilemmal carcinoma, trichilemmoma, eccrineporoma, clear cell hidradenoma	Red
	+ MPS-mucopolysaccharide (PAS after digestion)	Neutral MPS+, fungi (diastase resistance) *Lichenoid dermatoses (colloid body), Spitz nevi (Kamino body), lupus erythematosus (basement membrane), porphyrias (vascular walls), colloid milium, hyalinosis cutis et mucosal, cryoglobulinemia, malakoplakia, Gaucher's disease, Fabry's disease, reticulohistiocytosis, granular cell tumor, mammary and extramammary Paget's disease, dermatomycosis	Red
14.	Alcian blue*	Acid MPS (sulfated MPS at pH2.5 and 0.5, carboxylated MPS at pH 2.5 only)	Blue; nuclei: red
15.	Colloid iron*	Acid MPS	Blue
16.	Toluidine blue*	Acid MPS, nucleic acid mast cell granules *All three: Chondroid syringoma, metastatic adenocarcinoma, cutaneous mucinosis (pretibial myxedema), granuloma annulare (necrobiotic foci), alopecia mucinosa, malignant atrophic papulosis, digital mucous cyst, mucinous eccrine carcinoma, extramammary Paget's disease	Metachromatic purple
17.	Hyaluronidase	Hyaluronic acid	Hyaluronidase labile
18.	Mucicarmine	"Epithelial" mucin	Red

Amyloid Stains

19. Congo red	Amyloid	Red with apple-green birefringence
20. Crystal violet	Amyloid	Metachromatic red-purple
21. Thioflavin T	Amyloid	Silver blue or yellow fluorescence
22. Pagoda red	Amyloid	Orange with yellow-green birefringence

Lipid Stains

23. Oil Red O	Lipids	Red
24. Scarlet Red	Lipids	Red
25. Sudan Black B	Lipids	Black
	*All three: Xanthoma, necrobiosis lipoidica diabeticorum, rheumatoid nodules, histocytosis X, reticulocystiocytosis, Fabry's disease, Tangier disease, hyalinosis cutis et mucosal, sebaceous carcinoma	

Mast Cell Stains

26. Giemsa	Mast cell, acid MPS	Metachromatic purple
27. Leder (Naphthol AS-D chloroacetate esterase)	Eosinophil, Leishmania	Red
	Mast cell	Red
	*Both: Urticaria pigmentosa, telangiectasia macularis eruptiva perstans, systemic mastocytosis	

Micro-organisms Stains

28. Gram (Brown-Brenn)	Gram positive	Blue-purple
	Gram negative	Red
	*Bacterial infections, actinomycosis, nocardiosis, actinomycetoma, anthrax, malakoplakia	
29. Ziehl-Neelsen	Acid-fast bacilli	Red
	*Tuberculosis, Mycobacterium marinum, nocardiosis	
30. Fite	Acid-fast bacilli	Red
31. Warthin-Starry	Spirochetes, Donovan bodies	Black
32. Gomori's methenamine silver nitrate (GMS)	Fungus wall, Donovan bodies,	Black
	*Fungal infection, actinomycosis, rhinoscleroma (Frisch bacilli) norcardiosis	
33. Dieterle and Steiner	Spirochetes, bacillary angiomatosis	Black

B. Immunohistochemical Techniques

This technique is most useful in the diagnosis of tumors, and is also helpful with the identification of viruses.

Epithelial Tumors
1. AE1/AE3: Cytokeratin cocktail
2. CAM5.2: Sweat gland and sebaceous cells—low molecular weight keratin
3. EMA: Cutaneous adnexae, hair follicle and sebaceous glands
4. CEA: Eccrine and apocrine glands
5. GCDFP-15 (Brst II): Apocrine cells

Mesenchymal Markers
6. Vimentin: Intermediate filaments, from cytoskeletal material, fibroblasts, endothelial cells, smooth muscle, melanocyte, cartilage, lipocytes, macrophages, and myoepithelial cells.
7. Desmin: Intermediate filament, muscle
8. Actin and smooth muscle actin: Microfilaments in the contractile apparatus

Vascular Markers
9. Factor VIII related antigen: von Willebrand factor, Kaposi's sarcoma, angiosarcoma, endothelial cells
10. CD34: Human hematopoietic precursor cells, endothelial and mesenchymal cells, dermatofibrosarcoma protuberans
11. CD31: Endothelial cells
12. Ulex europeus: Endothelial cells, angiosarcoma, Kaposi's sarcoma

Neuroectodermal
13. S-100 Protein: Melanocyte, Langerhan's cells, sweat ducts, Schwann cells
14. HMB-45: Melanoma cells, some nevus cells

Neuroendocrine Markers
15. Synaptophysin: Neuroendocrine cells (Merkel cells)
16. Chromogranin: Merkel cells, carcinoid
17. Bombesin: Small cell carcinoma

Lymphoid and Hematopoietic Tissue
18. CD-45 (LCA): Hematopoietic cell lines
19. CD-30 (Ki-1): Lymphomatoid papulosis, ALCL, R-S cells
20. CD-68 (Kp-1): Macrophage
21. Factor XIIIa: Platelets, macrophage, dermal dendrocytes, dermatofibroma
22. CD1a(Leub): Langerhan's cells
23. Myeloperoxidase: Myeloid series
24. Lysozymes: Macrophages

T-Cell Markers
25. CD45 RO (UCHL1): CD8 and Cd4
26. CD43 (Leu 22): T and some B-cell lymphomas, Myeloid series
27. CD4: Helper T-cells
28. CD8: Suppressor T-cells
29. CD2, CD3, CD5, CD7: Pan-T cell markers

B-Cell Markers
30. CD45R (MTR2): B-cell lymphoma
31. CD20 (L26): pan-B cell marker
32. Kappa and Lambda: B-cells marker (light chain immunoglobulin)

Viral Markers
33. Herpes simplex I and II
34. Cytomegalovirus
35. Ebstein-Barr Virus

Undifferentiated Tumor Panel
Cytokeratin + : Squamous cell carcinoma
S-100 + : Melanoma
Vimentin + : Atypical fibroxanthoma
Actin and smooth muscle actin + : Leiomyosarcoma

Dermatofibroma vs Dermatofibrosarcoma Protuberans
Factor XIIIa + : Dermatofibroma
CD34: Dermatofibrosarcoma protuberans
Mycosis fungoides
CD3, CD4 +
CD5, CD7 -
T-Cell Lymphoma vs Myelogenous Leukemia
LCA and T-cell markers + : T-cell lymphoma
Myeloperoxidase and lysozyme +: Leukemia
Merkel Cell Tumor
Cytokeratin: AE1/3, CK20, CAM 5.2, MAK-6 (paranuclear dot)
Chromogranin, Synaptophysin
NSE, neurofilaments
Paget's Disease vs Bowen's Disease
Cytokeratin + : Bowen's disease
BRSTII, EMA, CEA, CAM 5.2 + : Paget's
Vascular Tumors
Factor VIII RA + : Vascular endothelial cells
CD34 + : Lymphatic endothelial cells
CD31(+) : Endothelial cells

23

C. Immunofluorescence

This technique is useful in vesiculobullous lesions, vasculitis and collagen vascular disease.

Vesiculobullous Dermatoses   ICS: squamous intercellular substance
                                 BMZ: basement membrane zone

Dermatosis	Immunoreactant	Site	Pattern
1. Pemphigus, all except	IgG	ICS	Lacelike
IgA pemphigus	IgA	ICS	Lacelike
paraneoplastic pemphigus	IgG	ICS	Lacelike
	IgG, C3	BMZ	Linear/granular
2. Bullous pemphigoid	IgG,C3	BMZ	Linear
3. Cicatricial pemphigoid	IgG,C3	BMZ	Linear
4. Herpes gestationis	C3	BMZ	Linear
5. Epidermolysis bullosa aquisita	IgG,C3	BMZ	Linear
6. Bullous SLE	IgG,C3	BMZ	Linear/granular
7. Dermatitis herpetiformis	IgA	BMZ	Granular
8. Linear IgA dermatosis	IgA	BMZ	Linear
9. Erythema multiforme	IgM,C3	BMZ	Granular
	IgM,C3	Vessels	Granular

Vasculitis
10. Henoch-Schonlein purpura: Deposition of IgA in capillaries

Antinuclear Antibodies

Pattern	Associated Antibodies	Associated Disorder
11. Homogenous	Anti-histone	SLE
	Anti-DNA	Drug-induced LE
12. RIM	Anti-DNA	SLE

13. Fine speckled	Anti-Sm	SLE, often with nephritis
	Anti-Ro/SSA	SLE, Sjogren's syndrome
	Anti-La/SSB	SCLE, neonatal lupus
	Anti-UIRNP	SLE, mixed CTD
	Anti-KU	SLE, scleroderma,
	myositis	
	Anti-topoisomerase 1 (SCL-70)	Scleroderma
14. Discrete speckled	Anti-centromere	CREST
15. Nucleolar	Anti-U3RNP	Scleroderma
	Anti-RNA polymerase 1	

D. Electron Microscopy

This technique is most useful in the diagnosis of vesiculobullous disorders in the neonate and is rarely necessary in some viral diseases.

1. Desmosome, melanosome
2. Keratinosome (Odland body)
3. Langerhan's cell (Birbeck granules) - Histiocytosis X
4. Merkel cell (specific membrane-bound granules)
5. Epidermolysis bullosa simplex: Bulla cavity above BMZ, in basal cells, as a result of degenerative, cytolytic changes in basal cells
6. Mast cell granules (urticaria pigmentosa)
7. Pemphigus vulgaris: Widening of squamous intercellular spaces
8. Bullous pemphigoid: Blister in lamina lucida
9. Herpes gestationes: Blister in lamila lucida
10. Varicella-Zoster virus: Herpes simplex
11. Molluscum contagiosum
12. Human papilloma virus (verruca)
13. Angiokeratoma corporis diffusum (Fabry's disease): Laminated lysosome
14. Granular cell tumor: Cytoplasmic granules (Lysosomes)
15. Collagen
16. Hunter disease: Lamellar body
17. Sezary syndrome: Large convoluted nuclei
18. Amyloid
19. Orf, Milker's nodule: Viral particles have oval shape with central dense core surrounded by thin electron dense layers

E. Polariscopic Examination

This technique mainly used for the detection of foreign material in specimens, lipid deposits.

1. Silica (granuloma)
2. Uric acid (gout) Fixation in alcohol is helpful
3. Amyloid (green birefringence after staining with alkaline congo red)
4. Fabry's disease (urine-maltese cross)
5. Trichothiodystrophy (hair shaft-alternating light and dark bands)
6. Cholesterol ester (xanthoma)
7. Wooden splinter, suture, starch granules

Appendix 1 - Spongiotic Dermatitis

Acute Spongiotic Dermatitis

- **Nummular eczema**
- **Contact dermatitis**
  papillary dermal edema;
  eosinophils
- **Dishydrotic eczema**
- **Photoallergic dermatitis**
- **Spongiotic drug eruption**
  eosinophils; deep infiltrate
- **Dermatophytosis**
  "sandwiched" ortho/para-
  keratosis; PMN in stratum
  corneum; PAS(+)
- **Erythema multiforme**
  dyskeratosis
- **Pityriasis rosea**
  mounds of parakeratosis;
  extrav RBC (but no
  hemosiderin)
- **Erythema annulare centrifugum**
- **Miliaria rubra**
  spongiotic microvesicle
  involving intraepidermal
  sweat duct
- **Id reaction**

- marked spongiosis with vesiculation
- perivascular lymphocytic infiltrate with
  papillary edema exocytosis of
  lymphocytes
- +/- impetiginization and/or scale crust
- NO acanthosis
- NO hyperkeratosis or parakeratosis

Subacute/Psoriasiform Dermatitis

- **Seborrheic dermatitis**
  parakeratotic shoulder at
  follicular opening; PMN in
  scale; yeast forms in str.
  corneum
- **Atopic dermatitis**
  keratosis pilaris (follicular
  plugging)
- **Contact dermatitis**
- **Nummular eczema**
- **Pityriasis rosea**
- **Stasis dermatitis**
- **Pityriasis lichenoides chronica**
  confluent parakeratosis; sparse
  lymphocytic interface pattern;
  melanophages

- acanthosis/psoriasiform hyperplasia
- hyperkeratosis with parakeratosis/
  scale
- focal spongiosis with microvesiculation

- **Dermatophytosis**
- **Guttate psoriasis**
    focal mounds of parakeratosis;
    exocytosis of PMNs; perivasc
    lymphs w/ extrav RBCs
- **Erythema annulare centrifugum**
    superficial perivasc lymphocytic
    infiltrate w/ eos ("cuffing");
    exocytosis; parakeratosis

Eosinophilic Spongiosis

- **Pemphigus vulgaris**
- **Bullous pemphigoid**
- **Pemphigus vegetans**
- **Herpes gestationis**
- **Allergic contact dermatitis**
- **Atopic dermatitis**
- **Arthropod bites**
- **Scabies**
- **Drug eruption**
- **Incontinentia pigmentosa**

Appendix 2 - Vacuolar Interface Dermatitis

With Epidermal Atrophy:

Erythema multiforme	dyskeratosis with surrounding lymphs (satellite cell necrosis); normal cornified layer; some degree of chronic inflammation in pap. dermis
Graft vs host disease	loss of keratinocyte polarity; dyskeratosis with surrounding lymphs (satellite cell necrosis); +/- involvement of hair follicle; + parakeratosis
Toxic epidermal necrolysis	full thickness epidermal necrosis w/ supepidermal bulla; minimal to no lymphocytic infiltrate; extravasated rbcs in bullae
Discoid lupus erythematosus	erythematous plaques with atrophy and alopecia; hyperkeratosis w/ follicular plugging; NO parakeratosis; periadnexal and interstitial lymphocytic infiltrate; interstitial mucin; vasodilatation with red cell extravasation
- Dermatomyositis **- Subacute cutaneous lupus erythematosus**	DM: periorbital violaceous rash or papules/plaques over bony prominences; SCLE: diffuse erythematous macules w/ patches of alopecia, subtle findings with mild perivascular and interstitial lymphocytic infiltrate (acanthosis poss.[Gottron's papules])

Lichen sclerosis et atrophicus	hyperkeratosis; homogenization of upper dermis; telangiectasia of vessels; band of lymphs in mid-upper reticular dermis
Poikiloderma atrophicans vasculare	erythema with slight scale; assoc w/ genodermatoses, early MF, or DM/LE; thin keratotic scale; edema of upper dermis with pigment incontinence & lymphohistiocytic infiltrate; dilated superficial capillaries
Chronic radiation dermatitis	hyperkeratosis w/ orthokeratosis; epidermal homogenization w/ dyskeratosis and basal vacuolization; pleomorphic dermal fibroblasts; atrophy/absence of adnexae

Without Epidermal Atrophy:

Fixed Drug Eruption	clusters of dyskeratotic keratinocytes; polymorphous infiltrate of superficial and deep vessels; prominent vacuolization and pigment incontinence/increased # melanophages; occ.eosinophils
Phototoxic drug eruption	sunburn reaction with vacuolization and apoptosis of keratinocytes; subepidermal edema; adnexa are unaffected; no eosinophils
- Subacute cutaneous lupus erythematosus	abrupt onset of symmetric, erythematous annular or psoriasis-like
- Neonatal lupus erythematosus	lesions of upper trunk/neck/arms; vacuolar change with colloid bodies; mild hyperkeratosis; more dense periadnexal lymphocytic infiltrate than acute LE; papillary dermal edema; focal extravasated rbcs
Pityriasis lichenoides chronica (and PLEVA)	recurrent brown-red papules covered with fine scale; resolves in 3-6 weeks; hyperkeratosis with confluent parakeratosis; spongiosis with exocytosis of rbcs & lymphs; occ. necrotic keratinocytes
Morbilliform viral exanthem	non-specific superficial perivascular lymphocytic infiltrate (rarely multinucleated keratinocytes)
Post inflammatory pigment alteration (i.e., erythema dyschomicum perstans)	EDP: macular hyperpigmented patches on trunk/arms/face of Latin Americans; pigment incontinence with dermal melanophages
Mycosis fungoides, patch stage	slight psoriasiform hyperplasia; lymphocytes often in linear array on epidermal side of BM zone; mild papillary dermal fibrosis

Appendix

| Paraneoplastic pemphigus | painful oral lesions with generalized eruption of papules/blisters; assoc. w/ underlying malignancy; +/- acantholysis |
| Secondary syphilis | psoriasiform epidermal hyperplasia with exocytosis of lymphocytes; superficial & deep perivascular lymphocytic infiltrate with plasma cells |

Appendix 3 - Lichenoid Infiltrate

| *hallmarks of lichenoid dermatitis:* | *(1) band-like infiltrate obscuring DEJ*
(2) vacuolar change of basal layer
(3) necrotic keratinocytes w/ colloid bodies
(4) melanophages |

With Interface Dermatitis:

Lichen planus	multiple pruritic papules/plaques of extensor surfaces; compact orthokeratosis (no parakeratosis); Civatte bodies; "saw-toothing" rete
Benign lichenoid keratosis	solitary nonpruritic papule/plaque; (+) parakeratosis; may see residual solar lentigo at edge of lesion; normal granular layer
Lichenoid drug eruption	numerous eosinophils; (+) parakeratosis; decreased granular layer; necrotic keratinocytes; mid and deep perivascular inflammation
Lichenoid lupus erythematosus	atrophic epidermis; hyperkeratosis with follicular plugging; deep perivascular and periadnexal infiltrate; dermal mucin deposits
Chronic GVHD	lichenoid pattern in *early* chronic GVHD (100+ days after BMT); "satellite-cell necrosis" (dykeratosis with adherent lymphs)
Lichen sclerosis et atrophicus	epidermal atrophy; hyperkeratosis; homogenization/sclerosis of papillary dermis; telangiectasia of superficial capillaries
Mycosis fungoides, plaque stage	slight psoriasiform hyperplasia of epidermis; fibrosis of papillary dermis; prominent epidermotropism (min. spongiosis) w/ atypia
PLEVA	extensive eruption of erythematous papules in young adults/children; wedge shaped infiltrate w/exocytosis of lymphs, rbcs; moderate keratinocyte necrosis; confluent parakeratotic scale with neutrophils
Lichen striatus	linear erythematous papular eruption (age 1-15); exocytosis of lymphs; focal parakeratosis; perieccrine lymphocytic infiltrate

Lichen nitidus	flesh colored papules (2-3mm) in children/young adults; dense infiltrate of lymphs and histiocytes in dermal papillae (ball & claw)

No Interface Changes:

Pigmented purpura (capillaritis) **(lichenoid dermatitis of Gougerot-Blum)**	crops of small red nonblanching macules which appear pigmented during resolution; dense perivascular lymphocytic infiltrate w/ extravasated RBCs; patchy parakeratosis
Lichen aureus	dense lymphohistiocytic infiltrate increased # dermal capillaries; focal exocytosis; scattered hemosiderin-laden macrophages
Inflammed seborrheic keratosis	typical features of SK with moderate lymphocytic infiltrate
Lichenoid actinic keratosis	nuclear atypia; variable number of plasma cells; hyperkeratosis with parakeratosis
Squamous cell carcinoma in-situ	full thickness cytologic atypia of epidermis often with immature appearance; plasma cells and/or eosinophils may be seen
Halo nevus	nests of nevus cells embedded in dense lymphocytic infiltrate
Melanoma	brisk lymphocytic response with overlying atypical melanocytic proliferation
Zoon's balanitis	glans penis of uncircumscribed male; epidermal atrophy; abundant plasma cells; dilated capillaries with RBC extravasation
Secondary syphilis	psoriasiform hyperplasia w/ basilar vacuolization; increased # plasma cells; spirochetes (silver stain +) found in minority of cases (epidermis/ wall of superficial vessels)

With Histiocytoid-Like Cells:

Langerhan's cell histiocytosis	often w/ ulceration & crusting; histiocytes mixed with lymphocytes and eosinophils
Mastocytosis	cuboidal to spindled mast cells (resembling nevus cells); highest density around vessels; may have increased epidermal pigmentation

Appendix 4 - Psoriasiform Hyperplasia/Dermatitis

Psoriasis	uniform elongation of rete ridges; confluent parakeratosis with decreased granular layer; neutrophils in stratum corneum (Munro's) and in spinous/granular layer (Kogoj)
• **pustular psoriasis** →	no epidermal hyperplasia; granular layer may be prominent
• **guttate psoriasis** →	minimal epidermal hyperplasia; diminished granular layer; focal parakeratosis; few PMNs; dilated blood vessels
Lichen simplex chronicus	pruritic lichenified plaques often involving lower legs, forearms, and posterior neck; acanthosis w/ irregular elongation of rete ridges; hypergranulosis; hyperkeratosis; papillary dermal fibrosis with vertically oriented collagen fibers
Prurigo nodularis	pruritic nodular lesion; histology similar to LSC; epidermal hyperplasia may be more irregular than LSC
Chronic spongiotic dermatitis	Features of LSC with added features:
• **atopic dermatitis** →	- eosinophils
• **seborrheic dermatitis** →	- follicular plugging and neutrophils in parakeratotic scale crust
• **nummular eczema** →	- focal mounding of parakeratosis with occasional eosinophils
• **contact dermatitis** →	- spongiotic vesicles (eos are <u>un</u>common in allergic contact)
Reiter's syndrome	young males; urethritis (neg. cultures); arthritis; conjunctivitis; marked hyperkeratosis w/ parakeratosis; PMNs in stratum corneum & spinous layer with may be large (resembling pustular psoriasis)
Pityriasis rosea	mounds of parakeratosis; focal spongiosis and exocytosis of lymphocytes; extravasation of RBCs; patchy lymphocytic infiltrate
Dermatophytosis / Candidiasis	aggregates of neutrophils within the parakeratotic scale; organisms typically found between an orthokeratotic layer and parakeratotic layer
Pityriasis rubra pilaris	follicular papules w/ perifollicular erythema coalescing into plaques w/ fine scale and distinct borders leaving islands of normal skin; hyperkeratosis with alternating orthokeratosis and parakeratosis ("checkerboard pattern"); follicular plugging w/ parakeratotic mounds at follicular ostia (shoulder parakeratosis)

Chondrodermatitis nodularis helicis	tender papule on rim of ear helix; epidermal hyperplasia with ulceration and scale-crust; fibrinoid change in dermis; inflammation of cartilage
Mycosis fungoides	epidermotropism of atypical lymphocytes w/ hyperchromatic, irregular nuclei with surrounding clear halo; these cells may line-up along DEJ; small aggregates of Pautrier's microabscesses
Actinic prurigo	photodermatitis in Native Americans with eczematous lesions in sun exposed areas; heavy perivascular lymphocytic infiltrate
Inflammatory linear verrucous epidermal nevus	majority are present by 5 years of age; grouped verrucous papules in linear arrangement; often on lower extremities; thickened and elongated rete; broad columns of parakeratosis alternating with compact orthokeratosis; no granular layer in areas of parakeratosis; exocytosis of neutrophils
Acrodermatitis enteropathica	secondary to zinc deficiency; eruptions of acral skin, scalp and perineum; pallor of superficial epidermis; extensive necrosis of individual keratinocytes; hyperkeratosis w/ parakeratosis
Glucagonoma syndrome	associated w/ islet cell tumors of pancreas (necrolytic migratory erythema); pallor of superficial epidermis with dyskeratotic keratinocytes; confluent parakeratosis
Secondary syphilis	superficial and deep perivascular infiltrate with lymphs and plasma cells; exocytosis of neutrophils; organisms in epidermis and vessels
Clear cell acanthoma	usually solitary on lower leg; well-demarcated areas of pale staining keratinocytes (except for basal layer, acrosyringia which are spared); absent granular layer; neutrophils with spongiosis
Bowen's disease	absent granular layer with overlying parakeratosis (+/- cutaneous horn); lack of maturation which may involve adnexal epithelium (esp. follicular infundibula); extensive solar elastosis

Appendix 5 - Superficial Perivascular Mononuclear Cell Infiltrate (w/o Epidermal Changes)

Tinea versicolor	minimal perivascular inflammation w/ slight hyperkeratosis; purple spores and short hyphae in stratum corneum on H&E
Post inflammatory pigment alteration	perivascular melanophages
Viral exanthem	nonspecific pattern with focal dyskeratosis & basal vacuolization

Morbilliform drug eruption	basal vacuolization; occ. eosinophils mixed w/ lymphs; extrav RBCs
Erythema annulare centrifugum	annular lesions of trunk/proximal extremities w/ fine trailing scale; early/superficial pattern will <u>not</u> show tight perivascular "cuffing"
Erythema marginatum	assoc. w/ rheumatic fever; PMNs are present with focal nuclear dust, but no fibrin deposition or RBC extravasation
Erythema gyratum repens	"wood grain" pattern of erythema; 25% assoc w/ internal malignancy; mild acanthosis and parakeratosis
Urticaria pigmentosa	seen in childhood; will resolve; mast cell infiltrate with blue-grey cytoplasm; occasional eosinophils; some mast cells are spindly; chloracetate esterase (= Leder stain) and giemsa stains [+]
Telangiectasia macularis eruptiva perstans (TMEP)	seen in adults; more persistent than UP; similar to UP but with fewer eosinophils and more dilated capillaries
With Eosinophils:	
Urticaria	sparse perivascular and interstitial infiltrate with eosinophils; mild dermal edema (spaces between collagen bundles)
Urticarial vasculitis	lesions persist > 24 hours; similar to urticaria but with leukocytoclasia and few extravasated RBCs
Dermal hypersensitivity reaction • **drug eruption** • **dermal contact hypersensitivity**	perivascular interstitial lymphocytic infiltrate with occasional eosinophils
Pruritic urticarial papules & plaques of pregnancy (PUPPP)	G_1 in 3rd trimester of pregnancy; same histology as urticaria with focal spongiosis
Herpes gestationis	more papillary dermal edema than PUPPP; focal keratinocyte necrosis leading to subepidermal blister with eosinophils
With Extravasated RBCs:	
Pityriasis rosea	focal parakeratotic scale; slight acanthosis; focal spongiosis
Pityriasis lichenoides chronica	confluent parakeratosis; focal keratinocyte necrosis; vacuolar change of basal keratinocytes (interface changes)
Pigmented purpura (capillaritis)	extravasated RBCs • Schamberg's = "cayenne pepper" rash • Majocchi's = annular lesions; telangiectasia

Stasis dermatitis	pruritic papules/plaques of lower legs; small vessel proliferation in upper dermis; hemosiderin deposition; fibrosis of reticular dermis
Kaposi's sarcoma, patch stage	subtle proliferation of irregular, jagged blood vessels around capillaries and/or adnexae which dissect between collagen bundles; bland appearing spindle cells forming vascular slit-like spaces; hemosiderin

With Neutrophils:

Cellulitis / erysipelas	dermal edema w/ interstitial infiltrate of lymphs and PMNs with some leukocytoclasia
Leukocytoclastic vasculitis	any 2 of the following: (1) PMNs with nuclear dust; (2) fibrin deposition in vessel wall; (3) extravasated RBCs

Appendix 6 - Superficial & Deep Perivascular Lymphocytic Infiltrate

Erythema annulare centrifugum	moderately intense, tight perivascular infiltrate ("coat sleave"); no epidermal changes or papillary dermal edema
Jessner's lymphocytic infiltrate	asymptomatic well demarcated red-brown papules/plaques on face or upper trunk; perivascular & periadnexal infiltrate w/o epidermal changes
Erythema chronicum migrans	assoc w/ Lyme disease; variable histology according to biopsy site: border→ normal epidermis and moderate perivasc/interstitial infiltrate center→ acanthosis with dermal hemorrhage and perivasc infiltrate
Polymorphous light eruption	clinical variability→small papules/vesicles/large plaques/prurigo-like lesions in sun exposed skin; moderate perivascular lymphocytic infiltrate with papillary dermal edema; +/- spongiotic microvesicles
PLEVA	epidermal dyskeratosis with moderate lymphocytic exocytosis into epidermis; abundant parakeratotic scale; RBC extravasation into dermis/epidermis; wedge shaped dermal lymphocytic infiltrate
Lupus erythematosus	patchy lymphocytic infiltrate with involvement of adnexae; epidermal changes (atrophy, vacuolar change, hyperkeratosis); RBC extravasat'n
Tumid lupus erythematosus	dermal form of LE without epidermal surface changes; patchy lymphocytic infiltrate with stromal mucin deposits

Arthropod reacation	may have dense inflammatory infiltrate with eosinophils
Papular urticaria	hypersensitivity reaction to bug bites with multiple papules; perivascular lymphocytic infiltrate with eosinophils; wedge shaped infiltrate which is broad superficially and tapers at the base
Perniosis	painful violaceous plaques on fingers/toes; moderate papillary dermal edema; moderate lymphocytic infiltrate which spares edematous papillary dermis
Cellulitis / erysipalis	dermal edema with diffuse infiltrate with numerous neutrophils; obvious leukocytoclasia
Inflammatory morphea	thickened collagen bundles; inflammatory infiltrate involving eccrine glands and extending into subcutaneous fat
Indeterminate leprosy	scattered pink/hypopigmented plaques; mononuclear infiltrate around vessels, adnexae, and nerves; no giant cells; only rare acid fast bacilli

Appendix 7 - Acantholytic Dermatitis

Suprabasilar Acantholysis *without* Dyskeratosis:

Pemphigus vulgaris	flaccid bullae of oral mucosa, face, scalp, trunk; suprabasal cleavage with tombstoning; (+) involvement of hair follicles
Pemphigus vegetans	verrucous epidermal hyperplasia; focal suprabasal acantholysis; eosinophilic spongiosis/pustules
Hailey-Hailey disease	hyperplastic epidermis with acantholysis half way up (dilapidated brick wall)
Transient acantholytic dyskeratosis (Grover's disease)	very focal (< 1 mm) involvement of epidermis; patterns include pemphigus vulgaris-like and Hailey-Hailey like
Paraneoplastic pemphigus	painful oral lesions in patients w/ underlying malignancy; suprabasal acantholysis with features of interface dermatitis

Suprabasilar Acantholysis *with* Dyskeratosis:

Definition of dyskeratosis
"corps ronds" are acantholytic keratinocytes w/ clumped keratohyalin granules in an eosinophilic cytoplasm w/ large nuclei with perinuclear halos "grains" are seed-shaped parakeratotic cells w/ pyknotic nuclei and dense eosinophilic cytoplasm

Darier's disease	suprabasal acantholysis w/ extensive dyskeratosis

Transient acantholytic dyskeratosis (Grover's disease)	very focal (<1 mm) involvement of epidermis; patterns include Darier's-like
Acantholytic actinic keratosis	often occurs on sides of face/periauricular; focal cytologic atypia in lower half of epidermis; NO corps ronds
Warty dyskeratoma	cup shaped invagination of epidermis in dermis with acantholysis and dyskeratosis similar to Darier's
Acantholytic squamous cell ca	nodules of SqCCa with central degeneration leaving cystic spaces lined by one or several layers of polygonal tumor cells; free floating dyskeratotic tumor cells are seen in the gland-like space

Subcorneal Acantholysis

Pemphigus foliaceous	subcorneal blister through granular layer (no orthokeratosis is present); minimal inflammation
Pemphigus erythematosus	erythematous papules/plaques in seborrheic areas mimicking lupus; identical histology as *P. foliaceous*
Staphylococcal scalded skin syndrome (SSSS)	superficial peeling of bright red skin (clinical DDX: TEN); sloughing of the stratum corneum w/ acantholytic keratinocytes; minimal inflammation

Miscellaneous

Epidermolytic hyperkeratosis	marked vacuolization of keratinocytes w/ enlarged keratohyalin granules; shrunken nuclei; psoriasiform hyperplasia w/ orthokeratosis
Herpes virus infection (herpes simplex/varicella/ zoster)	intraepidermal vesicle w/ acantholysis and reticular degeneration; nuclei w/ "ground glass" appearance and marginated chromatin; eosinophilic intranuclear inclusion w/ perinuclear halo; may involve hair follicle

Appendix 8 - Subepidermal Vesicular Dermatitis

With Minimal Inflammation:

Epidermolysis bullosa	
• **EB simplex** →	no scarring/milia; blister occurs through level of keratinocyte
• **junctional EB** →	no scarring/milia; blister occurs through basement membrane
• **dystrophic EB** →	(+) scarring/milia; blister occurs through level of papillary dermis
Epidermolysis bullosa acquisita	(+) scarring/milia; blister formation same level as dystrophic EB; immune complexes (C3/IgG) in *floor* of blister in salt split skin

Porphyria cutanea tarda	solar elastosis (sun exposed skin); festooning of papillary dermis into blister cavity; thickening of vessel walls in papillary dermis; segmented basement membrane structures in blister roof (caterpillar bodies)
Cell poor bullous pemphigoid	occasional eosinophils seen at dermal-epidermal junction; immune complexes (IgG/C3) in *roof* of salt split skin
Pseudoporphyria	secondary to dialysis or drug (i.e., furosamide); histology identical to PCT
Toxic epidermal necrolysis	full thickness epidermal necrosis with minimal inflammation
Bullosis diabeticorum	patients w/ severe diabetes and renal insufficiency; often involves hands/feet; thickened vessel walls in papillary dermis
Pressure induced bullae	patients in coma; necrosis of eccrine sweat gland w/ epidermal spongiosis and vesiculation
Bullous amyloidosis	often periorbital; clefting through amyloid deposits in upper dermis

With Lymphocytic/Mononuclear Infiltrate:

Erythema multiforme	focal to extensive epidermal keratinocyte necrosis w/ vacuolar degeneration; mild to moderate superficial lymphocytic infiltrate
GVHD, grade IV	epidermal keratinocyte necrosis; abundant lymphocytic infiltrate; individually necrotic cells in eccrine glands
Fixed drug eruption	interface dermatitis w/ vacuolar degeneration; keratinocyte necrosis w/ lymphocytic exocytosis; perivascular lymphocytic infiltrate in superficial and deep dermis; melanophages
Bullous lichen planus	band like infiltrate obscuring dermal-epidermal junction w/ colloid bodies; irregular acanthosis & hypergranulosis; epidermal separation/clefting
Bullous mastocytosis	affects infants and children; diffuse mast cell infiltrate often w/ "fried egg" cytological appearance; Giemsa/Leder stains (+)
Bullous fungal infection	mixed dermal infiltrate with lymphs, eosinophils, plasma cells; epidermis displays parakeratosis and spongiosis; PAS or GMS readily identifies
Bullous PMLE	sun exposed skin; esp. young females; epidermal spongiosis w/ papillary dermal edema; superficial and deep perivascular lymphocytic infiltrate

With Eosinophils:

Bullous pemphigoid	older patient pop.; dense inflammation (eosinophils) in papillary dermis and blister cavity w/ eosinophilic spongiosis; no epidermal necrosis
Herpes gestationis	pregnant females; same histology as bullous pemphigoid; Clinical DDX includes PUPPP: lymphocyte rich infiltrate w/o vesicle formation
Bullous drug eruption	mixed inflammation involving superficial and deep dermis with variable number of eosinophils
Bullous arthropod reaction	superficial and deep inflammation w/ eosinophils; may see punctum site in epidermis (discrete epidermal necrosis)
Epidermolysis bullosa acquisita	sometimes EBA has mixture of eosinophils and neutrophils
Linear IgA bullous dermatosis	histology similar to BP but increased number of neutrophils

With Neutrophils:

Dermatitis herpetiformis	grouped papulovesicles in young/middle aged adults; (+) pruritis; small subepidermal blisters w/ neutrophilic microabscesses in tips of papillary dermis; (increased # eos with older lesions); collection of neutrophilic infiltrate is usually discontinuous; superficial perivascular lymphocytic infiltrate
Linear IgA bullous dermatosis • **linear IgA disease (*adult*)** • **chronic bullous disease of childhood (*children*)**	histology similar to DH but with neutrophilic infiltrate more evenly distributed along dermal-epidermal junction; less tendency for papillary microabscess formation; more eosinophils than DH
Cicatricial pemphigoid	elderly; occular or oral mucosa; subepidermal split w/ mixed inflammation (neutrophils and/or lymphocytes); eosinophils less frequent; older lesions w/ granulation tissue and fibrosis; no acanthosis
Sweet's syndrome	marked papillary dermal edema with blister; dense band of neutrophils/lymphocytes in upper reticular dermis w/ leukocytoclasia
Bullous lupus erythematosus	DH-like pattern w/ subepidermal blister w/ neutrophils; dermal edema and mucin deposition; moderate perivascular inflammation; +/- LCV with leukocytoclasia and hemorrhage
Pustular drug eruption	papillary dermal collections of neutrophils; LCV with fibrinoid necrosis of vessel wall

Appendix 9 - Subcorneal/Intraepidermal Pustular Dermatitis

Impetigo/impetiginized spongiotic dermatitis	subcorneal and intraepidermal pustules with edema of upper dermis; may find gram (+) streptococci in pustule (impetigo only)
Bullous impetigo	acantholytic blister with less intense infiltrate of neutrophils (compared to impetigo)
Candidiasis	red papules/pustules in intertriginous areas; histology similar to impetigo; PAS(+) yeasts/ branching pseudohyphae in stratum corneum
Dermatophytosis	similar histology to candidiasis
Subcorneal pustular dermatosis (Sneddon-Wilkinson disease)	vesicles/pustules develop on erythematous skin; may be annular or serpiginous; tense subcorneal pustules with mild orthokeratosis; do not see single inflammatory cells in surrounding epidermis
IgA pemphigus	pruritic eruption of axilla, trunk, extremities in middle aged/older patients; similar histology as impetigo; DIF = intercellular IgA
Pustular psoriasis	can be localized to palms/soles or a generalized eruption; spongiform pustules in epidermis (+/- acanthosis); confluent parakeratosis; hypogranulosis
Reiter's syndrome	young males with urethritis, arthritis, & conjunctivitis; eruption involves palms/soles, glans penis, oral mucosa; same histology as pustular psoriasis
Pustular drug reaction	diffuse macular erythematous eruption w/ pustules occurring 3-10 days after initiation of therapy; neutrophilic spongiosis with or without intra-epidermal pustules; inflammation may involve acrosyringium/hair follicle; dermal perivascular infiltrate includes eosinophils
Acute pustular folliculitis	erythematous papular eruption; neutrophilic infiltrate may involve follicular ostium giving appearance of intraepidermal pustule
Transient neonatal pustular melanosis	vesiculopustules on trunk /diaper area/face; present at birth; subcorneal pustules with neutrophils and eosinophils; basal hypermelanosis in older lesions
Erythema toxicum neonatorum	macules/papules with pustules on face/trunk within first 2 days of life; subcorneal pustules w/ mainly eosinophils which may involve hair follicles
Acropustulosis of infancy	most often seen in black infants from birth-2 years; pruritic pustules; usually affecting extremities; non-specific subcorneal pustules; mild papillary dermal edema

Appendix 10 - Diffuse Neutrophilic Infiltrate

Infectious vasculitis	small vessel vasculitis w/ nuclear dust, fibrin deposition, extravasation of RBCs; neutrophilic infiltrate may have extensive dermal extension
Sweet's syndrome (acute febrile neutrophilic dermatosis)	abrupt onset of (painful) vesicles/pustules of face and extremities; marked papillary dermal edema; (+) nuclear dust; RBC extravasation
Pyoderma gangrenosum	begins as tender papules/folliculitis ulcerated lesion w/ bluish border; extensive PMN infiltrate w/ follicular involvement; deeper than Sweet's
Cellulitis / erysipelas	warm & tender indurated plaque [erysipelas→sharply demarcated with adv. edge]; moderate edema of reticular dermis; Gram stain usually (-)
Furuncle / carbuncle	large, tender nodules esp. on legs, buttocks, post. neck, axillae; *carbuncle* = 2+ furuncles; dense follicular neutrophilic infiltrate w/ follicular rupture
Erythema elevatum diutinum (EED)	red, soft nodules of extensor surface of extremities; *early*→LCV; *mature*→ fibrosis with vertical vessels and diffuse mixed infiltrate (mainly PMNs)
Behcet's disease	mucocutaneous lesions (oral/genital aphthae, uveitis, etc); *early*→ LCV; *late*→ lymphocytic or granulomatous vasculitis
Granuloma faciale	soft papules or plaques on face; dense dermal infiltrate (PMNs, eos, plasma cells); intact epidermis/adnexae (Grenz zone); nuclear dust w/ fibrin
Neutrophilic arthropod bite	polymorphous infiltrate with PMNs, eosinophils, and lymphocytes
Traumatic deep ulceration (factitial dermatitis)	histology similar to pyoderma gangrenosum
Suppurative folliculitis • **acne conglobata (trunk/neck)** • **acne keloidalis (occiput/neck)** • **hidradenitis suppurativa (axilla/groin)**	comedonal dilatation with dermal abscess with mixed inflammation/giant cells and dermal fibrosis; squamous lined sinus tracts/cyst formation; naked hair shafts (acne keloidalis)
Ruptured follicle /follicular cyst	mixed inflammation with PMNs and multinucleated giant cells (MNGC); may see keratin debris in abscess or in cytoplasm of MNCG
Bacillary angiomatosis	exuberant small vessel proliferation w/ plump endothelial cells; micro-abscess w/ nuclear dust; basophilic debris [bacilli→ Warthin-Starry (+)]

Suppurative Granulomas with Histiocytic Component

Blastomycosis	PEH with intraepidermal and dermal microabscesses w/ a granulo-matous infiltrate with MNGC; small spores w/ broad based budding
Pyoderma vegetans (blastomycosis-like pyoderma)	most commonly occurs in interiginous areas, face, legs; similar histology to blastomycosis (including PEH)
Chromomycosis	typically occurs on extremity; patient is otherwise well; PEH with variable # of pigmented organisms ("copper pennies")
Sporotrichosis	non-specific microabscesses in PEH/dermis; organisms only rarely seen on H&E (specific dx by culture)
Phaeohyphomycosis	slowly growing nodule/subQ cyst; fibrous encapsulation; may contain wooden splinter; budding spores and hyphae; Fontana-Masson (+)
Mycetoma	small nodule (most commonly on foot); may have sinus tract; microabscess with surrounding granulation tissue; PAS/GMS + hyphae
Nocardiosis	filamentous bacteria which may form sulfur granules; gram (+) beaded pattern; Fite stain (+)
Actinomycosis	often premandibular; neutrophilic abscess with sulfur granule composed of gram (+) filamentous bacteria; Fite stain (-)
Scrofuloderma	direct extension to the skin of an underlying TB infection (lymph node); abscess formation with tuberculoid granulomas; (+) acid-fast bacilli
Halogenodermas • **bromoderma (lower extremeties)** • **iododerma (face)** • **fluoroderma (neck and preauricular)**	non-specific, PEH with intraepidermal and dermal microabscess formation with eosinophils and extravasated RBCs

PEH = pseudoepitheliomatous hyperplasia

Appendix 11 - Granulomatous Dermatitis

Nodular, mononuclear cell granulomas (histiocytes predominate)

Sarcoidosis	20-40 yo; males>females; blacks > whites; epithelioid granulomas with Langhans type giant cells; no caseation (but may see focal fibrinoid necrosis); generally only few lymphocytes ("naked granulomas")
Granulomatous rosacea	face; perifollicular granulomatous infiltrate usually without caseation; may see follicular plugging

Foreign body giant cell reaction	resembles sarcoidosis with multinucleated giant cells attempting to phagocytize the substance, either endogenous (i.e., hair follicle contents) or exogenous (i.e., tattoo pigment, suture material, etc.)
Silica / beryllium / zirconium	histologically indistinguishable from sarcoid; silicosis will have refractile particles in granulomatous infiltrate upon polarization
Crohn's disease	indistinguishable from sarcoidosis
Granulomatous cheilitis (Melkersson-Rosenthal)	intermittent lip swelling; facial paralysis; fissured tongue; scattered small, naked sarcoidal tubercles often adjacent to dilated lymphatic channels
Tuberculoid leprosy	one (or few) well demarcated lesions; asymptomatic macules; similar to sarcoidosis with epithelioid granulomas; no grenz zone; often has nerve involvement; acid fast organisms hard to find
Lepromatous leprosy (Hansen's disease)	multiple cutaneous and mucosal lesions; foamy, histiocytic infiltrate with destruction of appendages (NO epithelioid granulomas); (+) grenz zone; many organisms on Fite stain
Lupus vulgaris	reactivation of prior infection; 90% involve head & neck w/ red-brown patches or deep "apple jelly" nodules; granulomatous infiltrate with minimal necrosis and many giant cells; (+) appendage destruction; only rare organisms
Tuberculid	hypersensitivity reaction (Id-like reaction); NO organisms seen
Secondary syphilis	older lesions of secondary syphilis may have sarcoidal granulomata in papular lesions
Granulomatous mycosis fungoides	non-caseating granulomatous inflammation in dermis along with more typical changes of mycosis fungoides in papillary dermis & epidermis

Diffuse, Mixed Cell Granulomas (Admixture of Histiocytes, Lymphs, and Neutrophils)

Ruptured hair follicle / cyst	cyst lining may not be present; may see cyst contents (laminated cornified cells) surrounded by histiocytes & PMNs; single cornified cells may be phagocytized by histiocytic giant cells
Deep fungal / mycobacterial:	suppurative granulomatous infiltrate often with pseudoepitheliomatous hyperplasia and:
• chromomycosis	"copper pennies" (6-12 µm) in clusters most often within giant cells
• coccidiomycosis	endospores (10-80 µm) either free or within giant cells; no hyphae
• blastomycosis	uniform, round spores (10 µm) with broad based budding

- **cryptococcus**   variably sized spores (3-15 μm) with narrow based budding; capsule is mucicarmine (+); also Fontana (+); usually no PEH
- **histoplasmosis**   very small spores (3 μm) found within histiocyte cytoplasm
- **sporotrichosis**   round/oval spores (4-6 μm) which are hard to find
- **actinomycosis**   sulfur granules consisting of gram (+) branched, filamentous bacteria; lightly acid fast
- **eumycetoma**   microabscess with septated hyphae surrounded by granulation tissue and fibrosis (located in deep dermis/soft tissue)
- **phaeohyphomycosis**   circumscribed, cyst-like inflammatory infiltrate with budding spores & hyphae (Fontana [+]); may see foreign body (i.e., splinter)

Palisaded/Caseating Granulomatous Dermatitis

Granuloma annulare	annular arrangement of flesh colored or pink papules; extremities of children and young adults; palisade of macrophages with central mucin (colloidal iron +) and degenerated collagen in *superficial* dermis; *multiple* small foci of degeneration in a single 4 mm punch bx
Subcutaneous GA (pseudorheumatoid nodule)	more often in children; lower extremities/buttocks/scalp/periorbital; similar to GA with central mucin but larger nodules located in subcutis
Necrobiosis lipoidica diabeticorum	two-thirds of patients have diabetes mellitus; atrophic yellow-brown patches/plaques of lower extremities (esp. pretibial); may be ulcerated; large foci of collagen degeneration (often only 1 focus in a 4 mm punch bx) located in mid to deep dermis with surrounded granulomas; scarring/fibrosis with telangiectasia and lymphocytic aggregates at periphery; usually NO mucin
Rheumatoid nodule	firm subcutaneous nodules overlying joints (i.e., elbow); bright red fibrinous necrobiosis with palisading histiocytic granulomas/giant cells; often obliterates adipose tissue; NO mucin
Actinic granuloma	looks like GA on sun exposed skin; central focus of basophilic solar elastosis in mid dermis surrounded by giant cells and histiocytes; multinucleated giant cells will phagocytize bits of elastic fibers; NO mucin
Lupus miliaris disseminatum facei	granulomatous rosacea with central caseation

Palisaded neutrophilic & granulomatous dermatitis	reaction pattern to primary vasculitis in skin; nodular GA-like infiltrate surrounding zones of degenerated collagen with scattered neutrophils and nuclear dust; later lesion resemble NLD with central eosinophilic fibrin deposition and scattered neutrophils
Necrobiotic xanthogranuloma	associated with IgG paraproteinemia; large areas of degeneration with lipid/cholesterol clefts; lymphocytes and plasma cells are often present
Scrofuloderma	direct extension of infection to skin surface from underlying lymph node or bone; caseating granulomas with lots of organisms
Gout	subcutaneous tophus of ear helix or overlying joint (elbow, hands, feet); pale staining, amorphous deposits (monosodium urate) with surrounding histiocytic/multinucleated giant cell response
Epithelioid sarcoma	young adult; affects extremities (esp. finger/hand/ forearm); slow growing, painless nodules or "indurated ulcer"; nodular growth pattern composed of epithelioid, eosinophilic cells with central necrosis; will find mitotic activitiy; may have intercellular deposition of hyalinized collagen; chronic inflammatory cells at periphery of tumor nodules (mimics inflammatory process); keratin/EMA (+); vimentin (+); CD34(+)

Entities *without* granulomatous inflammation:
- granuloma faciale
- eosinophilic granuloma
- pyogenic granuloma
- lethal midline granuloma

Appendix 12 - Sclerosing Dermatoses

Scleroderma	*early*→ superficial and deep lymphoplasmacytic infiltrate involving eccrine glands and subcutaneous fat; *late*→ (1) thickened, tightly packed, hyalinized collagen bundles which replace the lower dermis and subQ fat; (2) atrophy/loss of eccrine and follicular structures which are entrapped by sclerotic collagen; (3) preservation of elastic tissue (VVG+)
Morphea	morphologically indistinguishable from scleroderma; early morphea can have modest inflammatory infiltrate

Lichen sclerosis	white, smooth, polygonal flat papules/plaques with finely wrinkled surface; atrophic epidermis with vacuolar change; hyperkeratosis with follicular plugging; homogenous hyalinization of the papillary dermis with underlining band-like infiltrate of mononuclear cells
Sclerodermoid GVHD	most often seen 150-300 days following BMT; typical features of GVHD with vacuolated basal cells and epidermal atrophy; dermal fibrosis with homogenized collagen bundles similar to morphea which extend downward from the papillary dermis
Chronic radiodermatitis	atrophic epidermis with dysmaturation; marked dermal sclerosis with atrophic adnexal structures; bizarre radiation fibroblasts; obliterative small vessel changes
Scleredema	widely separated dermal collagen bundles with hyaluronic acid deposition; no hyalinization/fibrosis of dermis
Atrophoderma	depressed patches usually on trunk; atrophic epidermis; affected dermis is markedly reduced in thickness and exhibits a mild degree of sclerosis
Eosinophilic fasciitis (Shulman's syndrome)	history of hardening of skin; pain of extremities often following physical exertion; associated with hematologic disease (aplastic anemia, leukemia, etc.); Labs→ peripheral eosinophilia, hypergamma-globulinemia, increased ESR; thickening of fibrous septa of subcutaneous fat with lymphoplasmacytic infiltrate, occ. eosinophils→ eventually develops thickened, sclerotic bands of collagen of reticular dermis in parallel to fascia

E

F

G